S. De Vere Burr

Tunneling Under the Hudson River

being a description of the obstacles encountered, the experience gained and the plans finally adopted for rapid and economical prosecution of the work

S. De Vere Burr

Tunneling Under the Hudson River
being a description of the obstacles encountered, the experience gained and the plans finally adopted for rapid and economical prosecution of the work

ISBN/EAN: 9783337302320

Printed in Europe, USA, Canada, Australia, Japan

Cover: Foto ©Andreas Hilbeck / pixelio.de

More available books at **www.hansebooks.com**

TUNNELING

UNDER THE

HUDSON RIVER:

BEING

A DESCRIPTION OF THE OBSTACLES ENCOUNTERED, THE EXPERIENCE GAINED, THE SUCCESS ACHIEVED, AND THE PLANS FINALLY ADOPTED FOR RAPID AND ECONOMICAL PROSECUTION OF THE WORK.

BY

S. D. V. BURR, A.M.

Illustrated by Working Drawings of all Details.

TWENTY-SEVEN PLATES.

NEW YORK:
JOHN WILEY & SONS,
15 ASTOR PLACE.

1885.

PREFACE.

The general plans according to which some twenty-five hundred feet of the Hudson River tunnel have been built are new: in executing them novel methods have been introduced—great obstacles have been surmounted—most valuable experience has been gained. By their aid a daily average of three, four, and sometimes over five feet of tunnel was completed during shorter or longer periods of time; and when the heading was more than one thousand feet from the shore an average of three and one-third feet of tunnel was built each day for more than two hundred consecutive days—a performance in engineering which far surpasses anything heretofore accomplished through like material. So far some $1,100,000 have been expended. Considering the great size of the tunnel, the character of the material passed through, and the depth under water, this is one of the cheapest, if not the cheapest, specimens of submarine tunneling ever accomplished, and this in the face of the facts that many stages of the work were necessarily experimental, that there were several serious delays—one caused by the accident—and that there were constant hindrances to rapid and economical progress.

In the following pages the aim has been to plainly, concisely, and yet fully describe every stage of the work. Although the fundamental idea embodying the use of compressed air was always adhered to, yet those which might be termed auxiliary plans, or methods of doing certain pieces of work which presented features peculiarly their own, were modified in accordance with the ever-changing conditions. With the assistance of the drawings, which were so selected as to cover all the more important points, and which, taken as a whole, constitute a very complete and easily understood history of the tunnel, it is hoped that the character of the difficulties may be appreciated and the methods by which they were overcome fully comprehended.

The writer's opportunities for close and careful inspection have been most favorable from almost the beginning of the enterprise. It has been his duty, while attached to technical journals of this city, to examine and describe the operations at the tunnel. Visits were frequent, and the purpose was early formed of some day collecting and arranging the data obtained. The larger part of the articles then written appeared in *Engineering News*, and were made up of notes almost invariably personally obtained in the heading of the tunnel.

The writer wishes to acknowledge the kind courtesy of Mr. D. C. Haskin, manager of the company, who loaned the note-books and working-drawings of his office.

S. D. V. B.

NEW YORK, January, 1885.

CONTENTS.

CHAPTER VII.

CHAPTER VIII.

LIST OF ILLUSTRATIONS.

TUNNELING

UNDER THE

HUDSON RIVER.

CHAPTER I.

NECESSITY FOR A TUNNEL—HUDSON TUNNEL RAILROAD COMPANY—LOCATION, LENGTH, AND GRADE OF TUNNEL.

THE Hudson River, flowing between New York City and Jersey City, where seven great railroads carrying the bulk of freight from the West have their termini, has always most effectually delayed transportation, and rendered impossible the sure, and at the same time the quick, delivery of freight. The magnitude of the barrier thus presented may be conceived when we remember that New York City is not only the largest shipping port in the country, but is also the distributing point for freight between the South and West and the East, and from the East to centres West and South. We therefore find that the few hours lost in crossing the Hudson affect not only this immediate vicinity, but they also affect in a much more marked degree every point having business relations with New York, or through New York to points beyond.

The passenger side of this question is, perhaps, the more important, although the reduction by the use of a tunnel would amount to an average of but a few minutes; but these short periods consumed by ferriage become worthy of most careful consideration in this age of quick travel and obstinate competition.

There are but two plans of obviating this difficulty—building a bridge or a tunnel. The tremendous cost of the first, arising from many causes which will appear to the observer, places it out of the question. Projects for tunneling under this stream, which has a swift tidal current and a deep channel, have long been discussed, but by all of the old and established methods of submarine tunneling the cost has stood in the way, for the simple reason that, if built with private capital, the charges or toll would have to be so large, in order to pay an interest on the amount invested, as to make its use practically impossible. The benefit to be derived was not commensurate with the expenditure when it could only be obtained at the rate of a few inches per day; consequently it became evident that some cheaper plan must be brought forward before a roadway could be built under the river. That some of the established plans were perfectly feasible is most probable, and that they could only be prosecuted slowly and at excessive cost is most certain. The great size of the tunnel was no mean obstacle, for, in order to accommodate travel, it would have to be, if single, large enough for two tracks, since a single-track tunnel would not clear away the obstructions to transportation, and would cost much more than one-half the sum which one carrying a double track would. The question then resolved itself into one double-track single tunnel or two single-track parallel tunnels.

For the purpose of constructing a railroad through a tunnel under the Hudson between the two cities the Hudson Tunnel Railroad Company was organized, with a capital of $10,000,000, under the general railroad laws of New York and New Jersey. This corporation is independent of the railroads now existing, and was dependent solely upon its own financial ability to carry out the work it proposed.

In locating the Hudson River Tunnel three questions came prominently up for consideration. The various routes, upon any one of which the tunnel could have been begun, while differing essentially from each other in some points of minor importance, yet presented the same general features regarding the material to be passed through and the engineering difficulties to be surmounted. The river at this locality varies but little in width or depth, being a trifle over a mile across and sixty feet deep in the channel, and the shores and bottom do not change much either in composition or

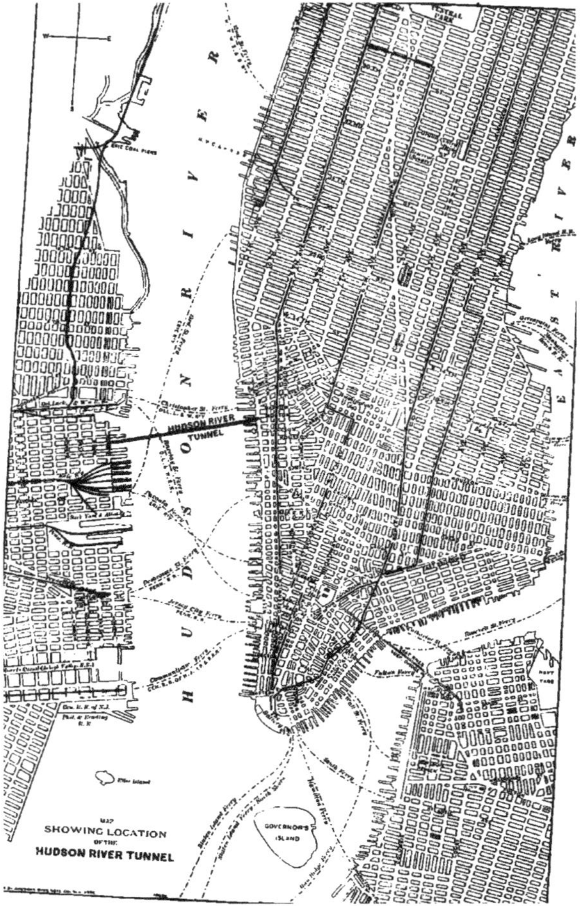
MAP
SHOWING LOCATION
OF THE
HUDSON RIVER TUNNEL
HUDSON RIVER
TUNNEL
H U D S O N R I V E R
E A S T R I V E R
GOVERNOR'S
ISLAND
CENTRAL PARK
ERIE COAL PIERS
NAVY YARD

contour between the extreme points at which the tunnel might have been located. The bed of the river consists of silt, which extends from the New Jersey or western side nearly or quite across the river; upon the New York side, near the shore, some rock is encountered, between which and the shore sand underlies the silt.

Owing to this peculiarity of formation the tunnel would be embedded for nearly its whole length in silt, no matter where its location might be; and, governed by these circumstances, the estimated cost of the work in each of the locations did not change much when confined to the tunnel proper. The approaches, of course, varied considerably with each location, and the selection depended directly upon the value of the land to be acquired at each terminus, thereby influencing the total cost of the work as a whole.

But the final and paramount question, upon which the real success of the scheme depended, was what might be termed the accessibility of each end, or, in other words, the convenience of using the stations and their adaptability to the requirements of trade. Upon the New York side this was governed primarily by the fact that it should be so situated as to be easily reached from the centre of the city, and yet the site should not be in a neighborhood in which the land damages would be so great as to be in reality prohibitory. Upon the other side of the river it had to be so located that the great railroads having their termini there could acquire access to it at a minimum cost.

After a careful examination of all the questions involved, the tunnel was finally located on a line extending easterly from Jersey Avenue (Jersey City), on Fifteenth Street, to Hudson Street, about 3,400 feet; from this point it curves five degrees northward to the New York City bulkhead line at the foot of Morton Street, about 5,500 feet, and thence again slightly southward about 4,000 feet to the eastern station. As will be seen by consulting the accompanying map, this line places the eastern end in one of the best locations in the city, and also furnishes the most convenient point upon which to concentrate the various railroad lines upon the New Jersey side with the least possible injury to existing interests. More than this, the line encounters as little, if not less, rock—which never comes nearer than 28 feet from the bottom of the river, and is wholly confined to a small knoll near the eastern shore—than any other, and insures plenty of head-room in silt. The composition of the material encountered along the entire line is shown very clearly in the plan and profile (Plate II.)

This plate also shows the grade of the tunnel. The slope from the western end is two feet in a hundred to a point about 1,600 feet from the shore, where the slope changes to one foot in one hundred; this continues to within 1,300 feet of the eastern shore, where, for a distance of 300 feet, it is one foot in one hundred. For 500 feet the grade is four in one hundred, and for 2,000 feet the grade is three feet in one hundred; thence to station the grade is moderate, conforming with the surface. The grade accords very nearly with the slope of the river-bed, and became necessary from the fact that it was decided that at no point should the crown of the arch approach nearer to the water than 15 feet. This was to insure safety in the work and to guard against any change which might take place in the future in the bed of the river.

Plate II.

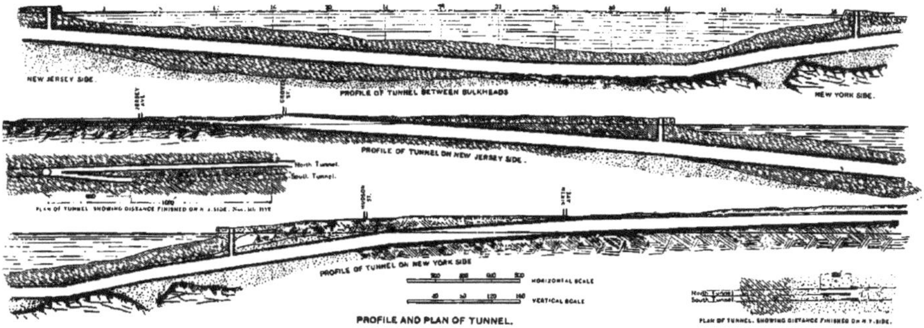

PROFILE AND PLAN OF TUNNEL.

CHAPTER II.

METHOD OF BUILDING—COMPRESSED AIR—FORM AND DIMENSIONS, AND SINKING, OF SHAFT—AIR-LOCK IN SHAFT—EFFECT OF AIR-PRESSURE UPON SILT—TEMPORARY ENTRANCE.

THE plan or method of carrying the work forward was by the use of compressed air, as applied in patents granted to Mr. D. C. Haskin, the president and manager of the company. This plan, briefly stated—for we shall describe its exact operation in detail further on—consisted in maintaining an air-pressure in the work about equal to the hydrostatic head. It was well known that if these two elements—the air-pressure inside and the water-pressure outside—could be so controlled as to maintain a constant and unchanging equilibrium, the material separating them need not be either of great size or strength. It was calculated that the silt, when in proper condition, would have tenacity or consistency enough to answer this object; and although it was known that this would only serve the purpose for a brief time—the exact duration could in no way be experimentally determined—it was thought that there would be ample time to place and secure the iron plates forming the exterior of the tunnel. It was known that, with the plates once in position and bolted together, the interior air-pressure would take the place of struts to keep them in position—a feature which would not only relieve the working space of encumbrance, but would give a much better support for the plates while the masonry was being put in. A plate 30 by 50 inches would be subjected, at a depth of 45 feet, to a pressure of 30,000 pounds. At a normal interior pressure this plate would have to be supported at several points in order to prevent bending, or, in case a single strut were employed, it would have to be made of great thickness and the strut would be of a size so large as to most seriously occupy the space. But with the use of compressed air at 20 pounds pressure per inch the plate would be supported at every point and its thickness would become a question of minor importance. It is hardly necessary to state that these things were all carefully considered before the work was begun. How well compressed air alone answered the purpose will be found as we advance with the undertaking.

After a year spent in taking soundings and boring on the line adopted, work on the tunnel was begun in November, 1874, it having been decided to sink a shaft on the New Jersey side near the river-line, and from the side of the shaft start the tunnel. The position of the shaft in relation to the tunnels is clearly indicated in the plan-view in Plate IV., and its position on the shore, 83 feet from the bulkhead-wall, is shown in Plate VI. It is circular in form, having an inside diameter of 30 feet, an outside diameter of 38 feet, the thickness of the wall at the bottom being 4 feet and at the top 2⅓ feet; the batter of the outer surface is one inch in three feet, or a total of 20 inches. It was sunk to a depth of 60 feet below the ground-surface, passing first through 9 feet of loose ash-filling, then 50 feet of silt, the bottom finally resting in sand. A shoe was first built in an excavation a few feet deep, and upon this the masonry (brick) was laid, the mass sinking as material was removed from under the shoe.

The shoe was made of 10 by 12-inch yellow pine, held together in ship-work style by drift-bolts. It had a cutting edge of boiler-iron, and was 4 feet high by 4 feet wide at the top, the section being triangular. Upon each of the east and west sides of the shaft there was built a false piece 24 feet high by 26 feet wide, having an elliptical form, and which was to be finally removed to make room for the approach and the tunnel. These sections were laid in common mortar—the rest of the shaft being laid in Rosendale cement—and during the sinking caused considerable trouble, as the false work in the river-side was forced in, by the pressure of the earth, to such an extent as to require interior bracing. The projection at the top amounted to 7 inches, decreasing to nothing at the bottom. The bracing consisted of heavy yellow-pine timbers formed into a hexagonal collar-brace; four sets, at varying heights, were put in, and, while they served the purpose admirably, they also acted as supports for the platforms afterward required.

Sinking of the shaft continued until December 15, 1874, when, after thirty days' work, the shoe was 14 feet below mean high water, when work was stopped by an injunction obtained by the Delaware,

Lackawanna & Western Railroad Company. On account of litigation work was not resumed until September, 1879.

Material was taken out by a hoisting-engine, the vibrations of which caused that part of the shaft nearest to it to sink faster than the remaining portions; and to overcome this difficulty the engine was moved about, when the settlement could be better controlled. On November 3 the shaft was in position, with the shoe 54 feet below mean high water, or 60 feet from the surface; the average rate of progress was one foot per day. During the sinking no difficulty was experienced in keeping the shaft free of water by the aid of an ordinary hand-pump; but as soon as the shoe entered the sand stratum water poured in at the rate of about 200 gallons per minute. While making the excavation two pumps were used alternately to keep the water down—an Andrews pump of 300 gallons' capacity per minute, and a Pulsometer of 100 gallons' capacity.

This movement displaced all the earth immediately adjacent to the shaft. The silt moved down the outside and under the shoe, and was followed by the ash-filling, the approximate positions, as determined later on, being indicated in Plates V. and VI. The appearance of the ashes, low down on the side of the shaft, was the direct cause of much trouble later on when the work was being prosecuted in this locality, since the ash mixing with the silt injured the latter's tenacity and made its behavior under air-pressure uncertain. The displacement was also aided by the sinking of the shaft itself, since, as it descended, the friction of its outer surface tended to dislodge the material and subsequently made the passage of the water down the side much easier. Building projecting courses about the shaft would not, in all probability, have obviated the difficulty; and if the work had been arrested for a time just before the shoe entered the sand stratum the water would not have been finally excluded from the bottom. If the latter course had been pursued it is doubtful if the shaft could have been started again on its course, so great is the clinging power of the silt when it has been allowed a little time to settle around an object. Piles driven along this shore move down easily when quickly driven, and when driven too far may be pulled up, no settlement taking place after the silt has had time to accommodate itself to the new conditions.

In order to lay the concrete bottom the water was confined as much as possible in covered chambers, a number of 4-inch iron pipes being laid (radiating) from a well having an open-slat wooden curb 4½ feet across at the bottom, 2 feet at the top, and 4 feet in height; this was lined with loose brick and placed at its full depth in a convenient situation in the bottom. That portion of the bottom least exposed to water was first covered with a foundation of securely-embedded dry stone from 12 to 18 inches deep, and then concrete was put on in layers from 8 to 12 inches deep. An adjoining section was then treated in the same way until the entire bottom had been put in and the water confined to the well-hole. The average thickness of the concrete was 2¾ feet—that at the centre being 2½ feet and the sides varying from 3 to 3¾ feet. The concrete was mixed as follows: One cement, two sand, two stone, and one gravel. Water, carrying sand with it, still came through the well, the filling of which with dry stone only stopped the sand to a small extent. Consequently, a 12-inch stand-pipe was carried up from the well for a distance of 25 feet, at which point a reducer was put on and a 4-inch length extended to above high-water level. The water in this pipe rose and fell with the tide, but not in the same proportion, nor did the level reach that of the river. The well-hole about the pipe was at last filled with concrete, which was mixed in bags and placed in the hole, which had been previously cleaned out and the sides dug away so as to form a cone. Other attempts to concrete this portion had been unsuccessful, as the water-pressure was sufficient to wash out the concrete as fast as it was put in; but by using bags the concrete was held until it had time to set, the outwardly and downwardly sloping sides preventing any dislodgment.

The air-lock (Plate IV.) was a cylindrical tube of ⅜-inch boiler-iron, built up of plates 3 feet wide and extending half-way round. Although it was 15 feet long by 6 feet in diameter, it proved to be admirably adapted for convenient use. In each end there was a door 3 feet wide by 4 feet high, and so hung upon strap-hinges 3 inches wide by ¾ inch thick, placed 10 inches above the bottom and the same below the top, and extending across the door, that it swung inward or toward the air-pressure. The door-frame was of 4-inch angle-iron, riveted to the head, which was stayed by five 2-inch braces that were 2 feet long and were riveted to the shell. The door was further strengthened by two strips of 4-inch T-iron placed horizontally. The doors and heads were each ½ an inch thick. There was a bull's-eye in each door and one in each head beside the door, made of 1-inch thick glass 9 inches in diameter, the exposed portion being 7½ inches across. Extending entirely through the air-lock were thirteen pipes varying from 1 to 4 inches in diameter and used for air and water supply, electric-light and telephone wires, blow-out pipes, etc. There was also a 6-inch pipe leading through the inner end of the air-lock and passing through the side of the lock to the bottom of the shaft. There was also a collar of angle-iron 6 inches wide and ½ an inch thick, securely fastened around the outside, against which to brace to secure the lock in position. The air-lock was used for the passage of men and materials, and was supposed to be the largest one ever built.

In the side of the shaft, at the top of the false piece previously mentioned, 29 feet below top of shaft, an opening was made to receive the air-lock. (In order to show the nature of this silt it may be well to state here that it gradually and regularly pushed its way through a small opening which had been made in the shaft, keeping the form of the hole until it projected a little way from the wall, when its own weight broke it off and it fell to the bottom, but still retained its shape.) This opening was not carried completely through the masonry, a thin shell of brick being left upon the inner side. The air-lock was then lowered down the shaft until it was opposite the opening, when, by means of hydraulic jacks braced against the opposite wall, it was forced through until it projected 4 inches beyond the outside of the shaft. Then the brick wall was fitted around the air-lock, being strengthened by a collar of yellow-pine timber, and securely braced from the opposite side of the shaft so as to prevent any change of position upon the air-pressure being put on from the front.

To remove the temporary door of boards which had been secured to the forward end of the lock, and which was held by the pressure of the earth against it, and begin the tunnel work was the next operation in order—one which was watched with absorbing interest, since the time had come to practically test the adaptability of silt for the work, as illustrated by its air-resisting properties. Men entered the lock, closed the rear door behind them, and after an air-pressure of 12 pounds to the square inch had been put on they removed the temporary door, when the silt was cleared slowly away until the iron door could be put in place and swung wide open. It was soon found that silt, when of the proper degree of moisture, was impervious to air, and the great problem, that the *silt would hold the air* and *the air would hold the silt*, was practically solved. At this time it was the intention to build a double-track single tunnel, 24 feet high by 26 feet wide in the clear, and the excavation was extended partly around the shaft, and on January 2, 1880, the chamber dug was 6 feet high, 15 feet wide, and 4 feet deep. No attempt was made to protect the silt from the air, which gradually forced back the water, minute holes first indicating an excessive dryness; these holes increased in size and developed into cracks, which extended up through the silt, in about four days, to the loose ash-filling that had been carried down by the shaft, as already mentioned, and through which the water flowed as they increased in size. Before the water found a passage the roof began to fall; first small, then large pieces of dry silt would become detached and drop. The top was kept in place four days by the air-pressure, but as the water had a disintegrating effect upon the silt, which seemed to mix with, and be held in suspension by, it, the excavation was refilled soon after the cracks appeared, the men retiring into the air-lock and closing the door.

A hole was now dug at the surface immediately above the inner end of the lock, 30 feet wide, 9 feet below high water, and extending 20 feet from the side of the shaft toward the river. Carefully spread over the bottom of this hole, and carried a short way up on the shaft, was a canvas, held down by heavy timbers, upon which the excavated material was replaced. This expedient was adopted in order to prevent the entrance of water into the work next to the shaft. While this work was being performed the plates for a temporary entrance leading to the inner lock-door had been made. This entrance (Plate IV.) was a tube 8 feet long by 6½ feet in diameter, composed of plates 2½ by 4 feet, with 3-inch angle-irons riveted along each edge. This tube lapped over the projecting part of the air lock, to which it was secured, and was made of iron ½ an inch thick; this amount of strength being considered requisite, since the tube was designed to prevent the air-lock door from ever becoming wedged should a caving of the roof take place.

Plate III.

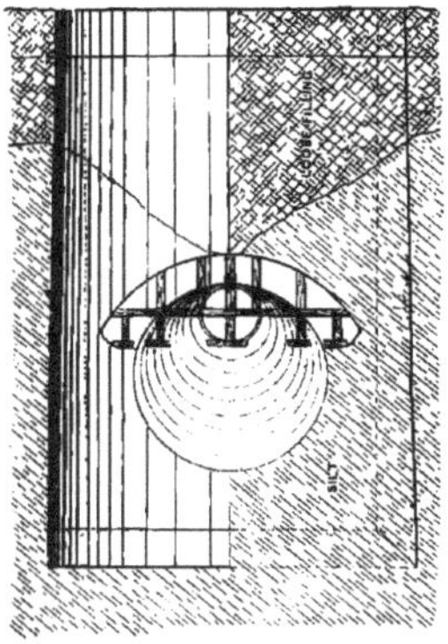

CROSS SECTION THROUGH TEMPORARY ENTRANCE
LOOKING TOWARD SHAFT FROM TUNNEL.

Plate IV.

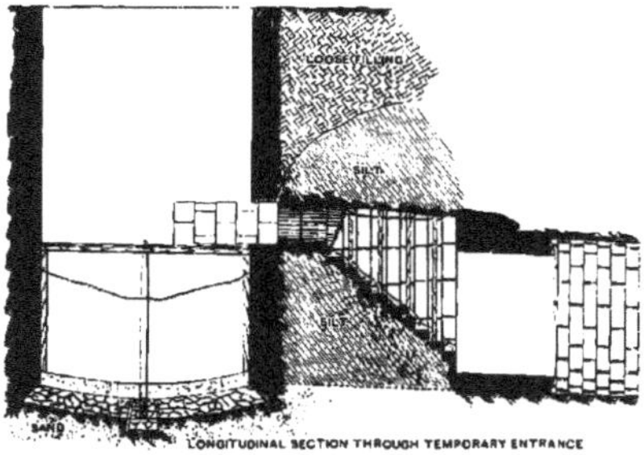

LONGITUDINAL SECTION THROUGH TEMPORARY ENTRANCE

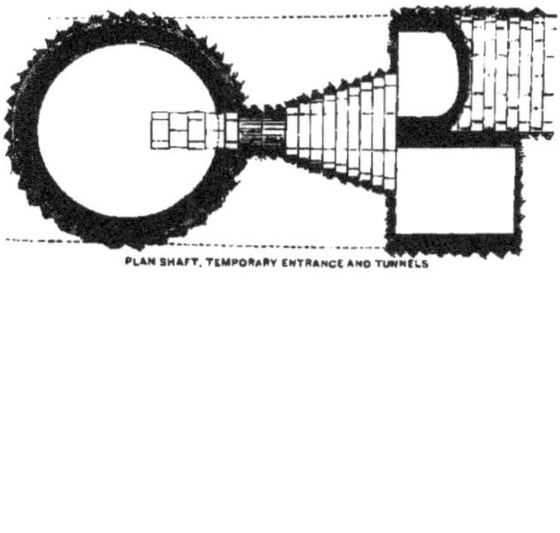

PLAN SHAFT, TEMPORARY ENTRANCE AND TUNNELS

Plate V.

PERSPECTIVE VIEW SHOWING CONDITION OF TEMPORARY ENTRANCE AT TIME OF ACCIDENT.

Plate III.

Plate V

From the end of the temporary entrance there was built a series of iron rings—to get away from the shaft into undisturbed silt—(Plates III., IV., and V.), each being 2 feet wide, 3/16 of an inch thick, and each being 18 inches larger than the one just preceding it. These rings were secured together at the top, which sloped so as to meet the proposed grade of the tunnel. The angular spaces formed by the lower portions of the rings were filled temporarily with concrete. The diameter of the largest ring was 20 feet. To place the plates forming these rings in position an excavation was made in the top adjoining the plate already in, when the centre upper plate of the next ring was put up and bolted to the finished work. Then an excavation was made at each end of this plate, to which the second and third plates were joined. As the circles were eccentric to each other, touching only at the top, the centre plates could only be held together by two or three bolts. The air-pressure was increased as the work advanced, and thus early it was learned that there could be no rule laid down by which to estimate the exact pressure needed. The correct amount fluctuated considerably, and to a certain degree was independent of the hydrostatic head; the density and composition of the material directly overhead exerted a direct influence upon the pressure, which could only be determined by closely watching the behavior of the earth under the pressure it was at the time exposed to, and change the air accordingly. After the completion of the eleven rings constituting the entrance the north tunnel was started, it having been decided to build two parallel single-track tunnels instead of one large one.

CHAPTER III.

STARTING TUNNELS—METHOD OF BUILDING—DISPOSITION OF EXCAVATED MATERIAL—BOLD PLAN ADOPTED IN BEGINNING SOUTH TUNNEL—DIFFICULTIES CAUSED BY OLD CRIB BULKHEAD—REMOVING TEMPORARY ENTRANCE—ACCIDENT—CONDITION OF WORK AT TIME OF ACCIDENT.

THE starting of this work was very difficult, requiring much care and watchfulness at every step. The north tunnel was the first one begun. The entrance, or connecting-chamber, as it was afterward more commonly called, being but 20 feet in diameter at its river end, would only cover the adjoining walls and not one-half of the arch and invert of each tunnel. To build the outer sections was, therefore, the hazardous part of the undertaking. But experience had already shown that, having a *firm foundation to which to attach an iron plate*, the work could be readily carried on in any direction or form. A space was excavated large enough to permit the insertion of a plate, which was secured by bolts to the plates of the last ring which served as a foundation to work from. Thus the work was prosecuted plate by plate until an iron band, or shell, of a size equal to the exterior of the tunnel, was constructed, in which the brick-work was laid. The vertical space between the north side of the entrance and the outer side of the north tunnel was securely bulkheaded, the plan being to build a certain length of each tunnel, and then return to replace the temporary work by uniting the tunnels and shaft. The air-pressure, about 18 pounds to the square inch, held the silt firmly in place while the rings were being made.

From this time the tunnel was built in the following way, which is very plainly shown in the drawing (Plate VI.) Silt was removed until the top centre plate could be put in and bolted to the one behind. Then a plate was put in at each side and bolted to the centre plate and to the ring. When this circle had been carried down the sides for some distance another was commenced and built in the

Plate VI.

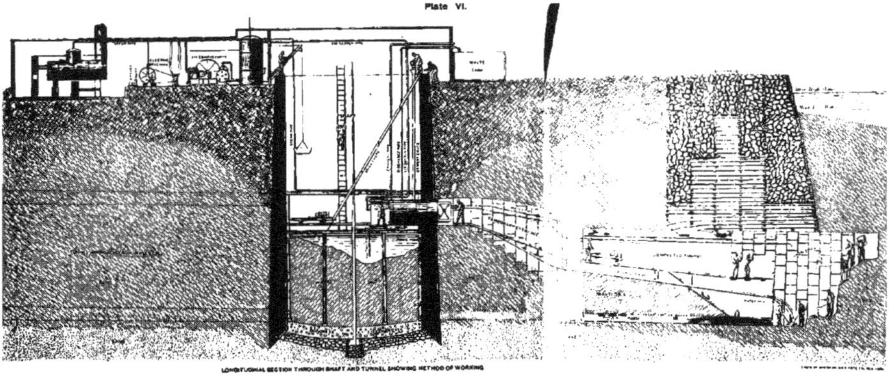

LONGITUDINAL SECTION THROUGH SHAFT AND TUNNEL SHOWING METHOD OF WORKING

same manner. When four rings of plates had been put up and thoroughly braced as additional protection in case of a reduction in the air-pressure, this iron-lined chamber was cleaned out and the masonry laid, thus completing a section of 10 feet. Both the iron-work and masonry were advanced in sections varying from 10 to 15 feet in length. During the first few weeks the rate of progress only averaged about 1 foot per day, but as the men became more familiar with the work, which was soon put in systematic shape, the rate increased to an average of nearly 5 feet of completed tunnel every 24 hours. The heading, which was cut into steps, as indicated in the drawing, was entirely exposed to the air-pressure, no attempt whatever being made to sheathe any part of it; but all digging to advance the heading was done carefully and without undue haste, since at any moment a less compact material, or "pocket," might be opened. It was the duty of some of the men to watch for leaks—in the temporary work at the entrance—which, if of some size, could be detected by the noise made by the out-rushing air; when small a candle passed close to the surface would show the leak by the flame being drawn in the hole by the escaping air. The leaks were easily controlled by the application of fresh silt.

About one-half of the excavated material—which was conveyed to the air-lock upon a car, the track being extended as the work advanced—was removed from the tunnel, the remainder being left in the finished work. The machinery and all the operations connected with the removal of earth and the introduction of supplies will be described in detail further on. At this time the silt was removed by being first mixed in a trough—shown to the right in Plate VI.—with water, and then blown through a 6-inch pipe by the air-pressure to the shaft, from which it was hoisted to the surface and conveyed to the low land just back of the works.

The shell was made of ¼-inch boiler-iron cut into plates 2½ feet wide and 3 and 6 feet long. On each of the four sides was a flange 3 inches wide; these flanges were pierced with holes at every 6 inches, through which each plate was bolted to the four plates around it. To give additional strength to the cylinder the joints were broken. The masonry was 2 feet thick, of hard-burned brick laid in cement. Four classes of men were employed upon the work, which was pushed forward night and day—miners, welders, laborers, and masons. The miners advanced the heading, the welders put up and united the plates, the laborers handled the diggings, and the masons placed the brick-work. The men worked in shifts of 8 hours each each shift consisting of 28 men, who were allowed half an hour's recess in which to come to the surface to eat their meal; but it was so arranged that one-half, or 14 men, remained on duty while the others went out.

The lines and grades of this portion of the north tunnel, which on July 1, 1880, had been finished for a distance of 281 feet from the shaft, were not good, and the shape was irregular. The reasons ascribed for this were that the plates, which were bent to a circular form, were joined to make elliptical rings, that the iron was too light, and that the plates were extended too far beyond the brick-work. At one time the plates on top were over 50 feet ahead of the masonry, the result being an almost imperceptible settlement, resulting in distortion. This was remedied by increasing the air-pressure and by not carrying the plates too far in advance of the finished work.

In order to prevent the escape of air through the pores of the brick, ordinary red lead paint was tried with but little success; much better results were obtained by a wash of pure cement put on in several layers.

The plan pursued in beginning the south tunnel was a bold one that proved to be rapid and successful in execution. The plates were inserted in small openings, as in the case of the north tunnel, but instead of removing all the material from the iron cylinder a centre core of silt was left. This core was about 5 feet long and a little less in size than the inside of the completed masonry. The men burrowed down each side and under this core, forming an annular chamber about 3 feet wide. As the excavation was carried down the plates were put in, and were supported by struts resting against the core, a cross-section of this part resembling a large wheel having an immense hub of earth, short, thick spokes, and a thin iron tire 5 feet in width. In justice to Mr. Anderson, the superintendent, it may be stated that he asked no one to venture under this core and put in the masonry; but, recognizing the fact that there was a possibility of the mass suddenly settling, he went under it himself and laid the first bricks in the invert. After this the masonry ring was quickly completed, yet it was some time before the core, which moved slowly and regularly, had settled down. The tunnel was then pushed forward until the advance plates struck the old wooden-crib bulkhead, when work on both tunnels was stopped, and operations were directed toward removing the temporary entrance, or connecting-chamber, as this was too weak to stand the increased pressure required at the heading, and permanently uniting the shaft and tunnels. At the end of the north tunnel a bulkhead was built of 4 by 6-inch timbers laid horizontally and closely together against the silt, with their ends resting against the ends of the brick-work ; behind these were vertical timbers of the same size, still back of which was another row of horizontal timbers, the last two rows being 3 or 4 feet apart. The whole was braced by struts supported in openings left in the masonry.

Tunneling through the wood and stone crib, which in some places projected half through the tunnel, was a difficult task. The ends of the piles had to be cut off and the horizontal timbers removed so as to clear the iron plates. It was through this old bulkhead that the water came at the time of the accident; it caused a great deal of trouble when recovering the bodies of the men and beginning anew.

The first work associated with the removal of the connecting-chamber was directed toward the last or largest ring. The two plates adjoining the centre one were taken down and the silt dug out, so that when the plates were re-inserted they were on the curve to be formed for the new work, the object being to construct a bridle, as it were, to cover both tunnels with one span or arch so as to leave a large chamber, as shown in Plate XI. In this way four rings were removed and the masonry built upon, as shown in Plate V. The plates in the roof of the remaining rings were then taken down, their place being supplied by the hood forming the crown of the new work. This hood reached from the completed work to the shaft, which it joined 3 feet above the air-lock, and then extended down each side and against the shaft as close as it could be well fitted, as shown in the cross-section (Plates III., IV., and V.) This roof was about 30 feet in width and was braced by timbers, as indicated in the drawings. The general condition of the work and the arrangement of the machinery at this time are shown in Plate

VI.—the north tunnel (not shown) had been completed for nearly 300 feet under the river.

At about 4.30 o'clock on the morning of July 21, 1880, when one-half of the men had returned to their work and the other half had started for the air-lock to pass out for their period of rest, a leak occurred, the air escaping up along the side of the shaft. The leak was probably caused by a plate not being properly secured against the shaft; but of this there is no definite knowledge. As the men were standing near the lock-door talking as to the best course to pursue the roof gave way, when the falling earth and plates so wedged the inner air-lock door—it was found open about 8 inches—that all efforts of the imprisoned men to open it were unavailing. The men were thus divided: eight were in the lock, six of whom had gone to stop the leak and two of whom had been out to lunch and returned, and twenty, including the assistant superintendent, Peter Woodland, were in the connecting-chamber near the door. When it was found absolutely impossible to provide any means of escape for the twenty men, the glass bull's-eye in the outer end of the lock was broken, and, the pressure being soon reduced to normal, the door was swung open and the eight men escaped. Upon the outer door being opened the water, coming from the old crib bulkhead, rushed through the lock and in a short time filled the shaft. How long a time elapsed between the discovery that the leak was beyond all control and the filling of the lock with water is not known with any degree of certainty; it must have been several minutes. The lock was amply large enough to have held the twenty-eight men. When work had been resumed the writer once passed through the lock with twenty men, and noted particularly that, with a little crowding, ten more could have been accommodated.

There are several points which must be mentioned in order that this stage of the work may be thoroughly understood. The material just above the lock was a mixture of silt and cinders, which had several times been disturbed, and which, while lacking the tenacious qualities of pure silt, was extremely treacherous. The cinder had gradually worked down until the dividing-line between it and the silt began about at the lock (this is plainly shown in Plate V., the letter A indicating the location of the blow-out), and extended upward and away from the shaft until it reached the undisturbed ground. Men were therefore appointed to watch for and stop leaks, which could be readily done with silt. The upper plates, forming a part of the permanent shell, were one-quarter of an inch thick, and were supported by timbers resting upon a foundation made in the silt floor. The hood thus formed had a large area for work of this kind; the surface was much too great to be carefully guarded, especially when a reduction of one or two pounds in the air was sufficient to produce deflection. The edges of the plates were not let into the masonry of the shaft, nor were they securely connected to it. The work had progressed so smoothly, and it being supposed that all danger had been passed, that, two or three days before the accident, several plates had been removed from the upper part of the temporary entrance, thereby leaving the door of the lock unprotected should anything fall from above.

Whether the blow-out first started as a small leak which was unperceived or neglected by the men in their haste to leave the tunnel, or whether it quickly developed into proportions placing it beyond their power to stop, will never be known. A blow-out in the tunnel could only occur in one of two ways: first, by a leak allowing the air to escape through the silt; second, by two leaks occurring simultaneously. In the first case the air would rush out until the interior pressure had been considerably reduced, when the water, taking the silt with it, would flow in; as the air became compressed it would again overcome the weight of the water and rush out. These movements—resembling the flow of water from an inverted bottle—would be repeated until the air had all been displaced, and would, of course, require more or less time. But in the second case the time necessary to effect the displacement would be much shortened, provided the conditions were such that one opening would permit the outward passage of air while the other admitted the water. The features governing this instance are so peculiar as to render it a most improbable, if not impossible, occurrence.

CHAPTER IV.

PLANS FOR RECOVERING THE BODIES AND OPENING THE WORK—OPEN CUT, COFFER-DAM, CAISSON—DESCRIPTION OF COFFER-DAM AND CAISSON—AIR-LOCKS IN CAISSON—SINKING CAISSON—OPENING OLD AIR-LOCK—FORCING SILT INTO SOUTH TUNNEL.

To open the work, so as to recover the bodies of the men and go on with the tunnel, was now the hard task. During the day and night following the accident all the pumping capacity that could be obtained was put in operation and an attempt made to pump out the shaft, but with no success. Divers, sent down the shaft, found that the air-lock was partly filled with silt, but they were unable to stop the flow of water. A few days later it was noticed that the water in the shaft rose and fell with the tide, clearly proving that there was a subterranean connection between it and the river. That the water was lowered in the shaft when sufficient pumping capacity had been obtained was due to the fact that it was only fed through the openings in the air-lock, which had been partially closed by débris when the water first rushed through.

The three methods which seemed to give the greatest hopes of success were: By an open cut; by building a roof within the shaft and forcing the water out by compressed air, and then entering by the old air-lock or by a new one built through the roof of the tunnel; and by sinking a caisson. Plans for a coffer-dam were immediately prepared, as it would be useful, if not essential, in either of the above methods. The reasons for this conclusion were that no open cut to the requisite depth of 35 feet could be sunk through that material without such a dam; that a roof or floor would be dangerous unless it could be embedded in the silt underlying the loose filling which covered the top of the connecting-chamber; and that, in the event of the caisson becoming necessary, it would be decidedly better to have as much excavation as possible in an open cut. Borings made after the caving showed that the loose filling was about 20 feet deep, and, with only the loose stone-crib bulkhead along the river-front as a protection, a deep excavation without the dam was an impossibility.

The coffer-dam was therefore begun. It was 46 feet square (Plate VII.), and was so built as to embrace a portion of the shaft which entered the inside wall to a distance of 11 feet, and also to extend about 5 feet east of the west end of the completed tunnel. The guide-piles were 12 inches square and the sheet-piles were 6 inches thick by 12 inches wide. The seams were calked to prevent the entrance of water. The timbers were of yellow pine, and were driven to a depth of 40 feet, great care being exercised in driving those forming the east wall, lest they should injure the roof of the tunnel. Excavation had been carried on simultaneously with work on the dam, and at 10 feet below the top a tier of beams was placed in position. As the depth became greater water began to interfere with the work, and so increased in volume that three pumps, throwing about 8,000 gallons per minute, could no longer control it. A second tier of beams 15 feet below the top of the dam was put in. The impossibility of handling the water compelled the cessation of further operations in an open cut.

The next method was to sink a caisson. A correct idea of the shape of the caisson and the relative dimensions would be obtained if a piece of stove-pipe, having a length a little greater than one-half its diameter, were cut by a plane passing through it parallel to, and a little above, its axis, the ends of the smaller segment so formed closed, and the whole covered by a nearly rectangular box.

The caisson was 41½ feet by 24 feet 10 inches outside at the bottom, 22 feet high outside, and the sides had a batter of 2 feet. The interior chamber was 40½ feet in diameter and 17 feet from the crown of the arch to the centre of the chord; the radius of the arch was 20⅓ feet.

The arch of the caisson was composed of yellow-pine timbers 6 feet long, 10 inches thick, and 12 inches wide. These were broken-jointed and bolted together with four ⅝-inch bolts through each. Placed in the interior was a lagging of planks 4 by 10 inches, which, after being well calked, was covered with sheets of lead and asphalt to render it impervious to water. The ends of the caisson were made of double timbers running in different directions, which were held firmly against the horizontal braces of the inside, and the roof by 17 bolted iron rods. Rods also extended in a transverse direction and bolted to the exterior to prevent spreading. The spaces between the top of the arch and the corners of the box were entirely filled with concrete to give additional weight and strength. An open box about 12 feet deep was built on top of the caisson, and was filled with the excavated earth in order to obtain weight to sink it. The construction of the caisson is clearly illustrated in Plates VII. and VIII.; it was designed by Mr. Anderson.

Through the roof of the caisson extended two air-locks—the smaller was 2½ feet in diameter and was for supplies; the other was 5 feet in diameter at the lower end and 6 feet at the top. These were built of plates of boiler-iron, which were added section by section as the caisson descended.

The larger lock was placed near, and about in the centre of, the western wall of the caisson. The section 5 feet in diameter extended from the roof nearly to the surface of the ground, where it expanded to 6 feet in diameter. On top was formed a circular chamber 6 feet high by 6 feet across, and attached to one side of which was a circular elbow 4 feet long by 5 feet across. In this portion three circular doors, 2 feet 10 inches in diameter in the clear, were hung—one at the outer end of the elbow, one between the elbow and the inner chamber, and one at the lower side of the chamber on top of the tube. The object in having three doors so located was to provide a large lock to be used when changing shifts or in case a blow-out should drive many men up the shaft. The smaller lock was used when three or four entered, and by it the passage could be made without losing so much air. The corners were stayed as shown in Plate IX. Leading from the lock down to the caisson-chamber were two ladders. By this method of construction the men were afforded a safe retreat in case of accident.

During the lowering of the caisson the water was expelled by air, the pressure being increased as the depth became greater. Each of the north and south sides of the caisson was upheld by three wrought-iron shoes 6 inches wide, 1 inch thick, and bent so as to clasp the under edge of the side. From these the screw suspension-rods, 3 inches in diameter, extended to three pairs of heavy timbers which projected about 1 foot over the sides of the coffer-dam and ran back upon the surface of the ground 30 feet, the ends being loaded with rails. This arrangement, shown in plan in Plate VII., gave ample strength to sustain the weight of the entire structure and prevented a too rapid descent. The caisson was lowered by digging away under the edges, and then, after having reduced the air-pressure, the suspension-rods were lengthened. In October, 1880, the caisson reached the requisite depth, 42 feet below the top of the shaft, or 38 feet below mean high tide. As the old connecting-chamber was cleaned out the bodies of the twenty men who had perished in the July accident were recovered.

When a trench had been dug under the edges of the caisson and the rods lengthened the air-pressure was reduced. The air served as a cushion which supported the greater part of the weight of the structure and its load, the side-rods regulating the rapidity of the downward motion and guiding the caisson. The amount of the reduction necessary to effect a movement varied with the friction of the sides through the silt, the clinging power of which was tremendous; the longer the time in which the silt had to settle, the greater the friction. The caisson weighed 460 tons; the box of earth on top weighed 350 tons; the locks and connections 45 tons; iron rails and bricks that were placed on top weighed 250 tons—making a total weight of 1,105 tons. The caisson was of sufficient strength to have withstood the pressure of the earth surrounding it, even if all the air had escaped. When the vertical struts were being placed in position one of the last acts was to drive small wedges of soft pine between them and the roof, designed to serve as indicators in regard to a yielding to strains. These were frequently examined, but at no time could the slightest change be observed. The weak parts of the caisson were the two ends, which were made of horizontal and vertical timbers resting against eight struts 1 foot square, extending the length of the caisson and braced by 3 transverse beams and 9 vertical struts.

A calculation was published by the writer at that time to show the greatest strain that could possibly come upon the braces. The area of the circular segment of each end was 514.4 square feet. This was calculated from the inner dimensions of the arch. Considering this as being at an average depth of 35 feet, and the mixture of silt and water as weighing 90 pounds per cubic foot, if not sustained by compact silt—and it was not probable that it would receive much, if any, support from this, since it had not settled firmly—there would be a pressure exerted upon the segment of 1,620,293 pounds, or 810 tons, or 3,150 pounds per square foot. The lower central strut supported an area of about 42 square feet, and, consequently, a combined pressure of 132,300 pounds. The two adjoining struts had to sustain an area of at least 77 square feet and a pressure of 242,550 pounds. This strain would have been by no means excessive and would have left a wide margin of safety.

After the sinking of the caisson the great problem was to form a connection between it and the shaft, and also between it and the completed portion of the tunnels. The case created much interest among engineers, many of whom criticised the plans very freely and had no hesitancy in predicting that work would never be resumed unless the methods were changed.

Operations were directed toward opening the old air-lock, the door of which was but about 2 feet from the west side of the caisson. To do this a rectangular opening, just large enough to admit a plate, was cut in the side of the caisson next the shaft, and the plate telescoped on to the projecting part of the air-lock. A second opening was cut next to the first, and a second plate inserted; this was continued until a perfect circle had been formed, when the centre piece was removed and the plates securely bolted to each other and to the side, and the interstices filled with silt. Air was then admitted to the lock, the outer door having been closed and the plate removed which had wedged the inner door at the time of the accident. Opening this lock greatly facilitated subsequent movements, since it was both easier and quicker to lower supplies down the shaft, and pass them through the old lock, than it was to admit them through the vertical lock.

It was decided to extend the two tunnels to the shaft, and not finish the upper half, but leave the whole as one large chamber uniting the two. The lower portion of the tunnel had been so far finished that the caisson extended about 8 feet over it, and the sides slanted until the top was about 4 feet from the outside. Building the invert, which was also to serve the purpose of invert for each tunnel, required great skill and extreme caution. A hole 4 feet square was started in the southeast corner, and as it

Plate VII.

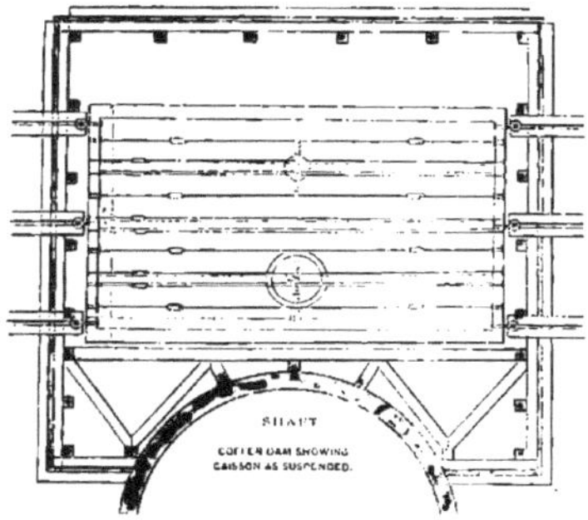

Plate VIII.

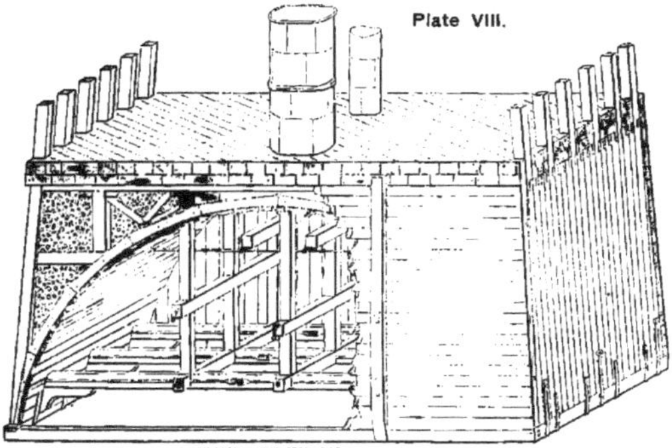

CAISSON. – NEW JERSEY SIDE.

Plate IX.

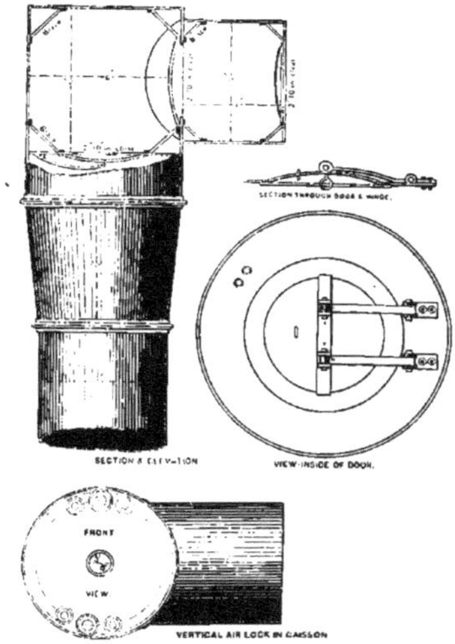

VERTICAL AIR LOCK IN CAISSON

was sunk the sides were planked and braced to secure the air. After the bottom had been reached, 8 feet below the shoe, the brick-work was started and built up on plates previously set, until that part of the caisson rested upon a secure foundation. The same was done at the northeast corner. A little at a time, and with great care, these plates were carried along the entire north and south sides, thereby forming a secure wall to resist the air-pressure. They were also carried over until the east side rested on masonry, except about 10 or 12 feet in the centre. A bulkhead was gradually built under this side down to the bottom, when the dividing wall, or centre column, was extended and the work of excavating the centre begun.

Plates 6 feet long by 2 feet wide had been telescoped, in the same manner as already described in connecting the old air-lock, from the eastern side of the caisson over the top of the tunnel, and securely bolted. Upon the western side a bulkhead had been built to the bottom, and heavy beams extended from side to side. It was early found necessary to bulkhead the openings to the two tunnels, so that the masonry in the caisson might be completed.

The caisson was not sunk so as to evenly cover the two tunnels, as will be seen in Plate XI. One shoe was in line with the exterior of one tunnel, while the other was about 3 feet from the tunnel. The masonry was built out to reach the sides, and projections were formed for the shoes to rest upon. A masonry arch was built 3 feet thick inside the arch of the caisson, and between this arch and the top of the tunnel was placed a brick bulkhead. This construction is very plainly indicated in the above-named drawing. The opposite or west side of the caisson was also bricked up from the bottom of the tunnel to the arch; this wall was afterward removed when the tunnels were joined to the shaft. Thus the caisson was converted into a large working-chamber, connected with the outer world by three air-locks, and wholly enclosed by masonry.

All the interior bracing of the caisson that did not interfere with the work was left in, and as fast as the excavations were made both vertical and horizontal struts were put in. A working-platform was erected, upon which were the water-tanks, cement-trough, bricks, etc. Rosendale cement, in the proportion of one cement to one and one-half sand, was used; it was mixed dry outside and brought in in bags.

CHAPTER V.

RESUMING WORK—ILLUMINATION—CONDITION OF ATMOSPHERE—AIR-COMPRESSORS—STEAM SUPPLY—TELEPHONE—SUPPLYING MATERIAL—REMOVING EXCAVATED MATERIAL.

THE caisson having been made perfectly secure, the next aim was to resume work on the tunnels. It was well known that a grave obstacle would be presented when the attempt was made to open the south tunnel, which had been completed but a short distance from the end of the connecting-chamber, and a few feet beyond which was the crib-work of the docks, through which water had ebbed and flowed since the day of the accident. That section adjacent to the caisson bulkhead had been compactly filled with silt, as just described, but beyond that no reliance could be placed upon compressed air, as the consistency of the silt which had been artificially thrown in decreased rapidly.

A hole 6 feet in diameter was cut through near the top of the bulkhead, and flanged iron plates 2 feet wide were put in and bolted. This small tunnel was constructed in a manner precisely similar to that by which the large tunnel was advanced—plate by plate. When the soft silt was met a piston was loosely fitted in the pilot. In this was placed a pipe, 3½ inches in diameter, through which silt could be forced. The piston-rod extended across the caisson to the opposite wall, and was operated by two hydraulic jacks. By this system of ramming the desired compactness was obtained at the head of the pilot.

When the crib-work was reached it was found that air would occasionally escape through the spaces between the loose filling. The application of a ball of silt would stop a small leak, but in large openings a bag of cement was forced up, when the air-pressure would hold it in position.

Practically the work was now in the same state as at the time of the accident, with the important exception that the temporary entrance had been replaced by a substantial chamber of masonry and additional facilities had been made for entering the tunnel.

We have not attempted to describe any of the machinery or the mode of getting in supplies and removing waste material; it was thought better to delay this until the work had been reopened.

For general illumination of the tunnel the electric light, which was early introduced and continued through all the operations, proved most satisfactory, one lamp lighting up very brilliantly a distance of from 100 to 150 feet. Its power depended upon the clearness of the atmosphere in the work; the haziness, which prevailed to a certain extent all the time, coupled with the almost total absence of reflecting surfaces, compelled the adoption of candles, which were used by the men to light all spaces which were shadowed. The masons had to use candles even when working directly beneath or alongside of the electric light when the rays from the latter were intercepted by a brace or other obstruction. Only the arc light was tried.

The dynamo and lamps were from the United States Electric Lighting Company; in the first lamps the carbons were ⅝ of an inch in diameter by 12 inches long. Ordinary coach candles 7½ inches long, 1¼ inches in diameter, and weighing 5 ounces, were used. These candles burned at the same rate inside as out of the tunnel.

As the heading was advanced the lamps were distributed along the tunnel, the object being to light the working-chamber where the locks were and the heading as brightly as possible, and illuminating the intervening space just enough for the workmen to easily find their way. Arising from the electric lamps and the candles there was much carbon-dust floating in the air; this collected in the nose, but was freely expelled by a vigorous blowing; it caused no ill effects whatever.

Telephone-wires connected the working-chamber (Plate XI.) with the office. The Bell telephone and Blake transmitter were used, the sounds being very distinct at each end of the line, but owing to the noise at the inner end it was found expedient to place the instrument in a small closet.

The success of the work at every step depended upon there being a constant and ample supply of air—a supply which had to be sufficient not only to overcome the loss due to small leaks, but to keep up the required pressure in case a large leak occurred. It was not possible to make and keep all the joints perfectly air-tight, and there was always a more or less quantity of air making its way out through the silt; besides this, the continuous use of the locks and the method of removing the waste silt—which will be described shortly—required a large amount of air. Within certain limits the free use of the air in the tunnel was not considered as an obstacle, since, by its introduction at the heading and its escape at the rear end of the work through the blow-out pipe and the locks, a circulation was kept up which insured a pure atmosphere in all parts.

At the time of which we are writing two air-compressors, used alternately, formed a part of the plant—a double-acting Clayton compressor, having two air and two steam cylinders, each 10 inches in diameter, and a stroke of 13 inches; a double-acting Ingersoll rock-drill compressor, having single cylinders, the diameter of the steam-cylinder being 10 inches, the air-cylinder 12 inches, and the stroke 12 inches. In actual practice the first machine compressed 1.65 cubic feet of air at each revolution; the second 1.05 feet at each revolution. A short time before the accident, when the needed pressure was 18 pounds per square inch, there were 83,000 cubic feet of normal air delivered into the tunnel every twenty-four hours; this giving about 150 cubic feet of normal air per hour to each man at work.

Air from the compressors was delivered to an air-reservoir built of boiler-iron, located near the shaft, and which was provided with a mercury-gauge to indicate the pressure. The delivery-pipe led from the reservoir, down the shaft, through the air-lock, and along the completed work to the heading. All the air was washed twice —once as it was drawn into the cylinders of the compressors, and again as it passed through water into the reservoir.

A steam-pipe led from the boilers to a force-pump near the old air-lock, supplying first a donkey-pump, then the electric-light machine and the air-compressors. After the working-chamber had been completed both a steam and air supply pipe were passed through the vertical lock, the former to run the engine used to draw up the cars.

In Plate VI. is shown a brick-chute, by means of which bricks were delivered on the platform in the shaft. It was a wooden box placed in an inclined position, and having upon the under surface of the upper side hinged leaves which retarded, without stopping, the descent of the brick, thereby insuring its safe arrival at the bottom. The cement was also sent through a chute to the platform. Both the brick, and cement which was mixed with the sand and tied up in bags, were loaded on cars running upon a narrow track which extended from the shaft through the air-lock down to the heading. A short, removable section of track fitted over the sill of each lock-door. The loaded car was run into the lock, the piece of track removed, the door closed, and the air from the tunnel admitted; then the inner door was opened, the track made continuous, and the car run out. In order that one car might be loading during the journey of the other there were two tracks laid in the shaft; the shifting of the cars was accomplished by the aid of a sliding platform. The general style of dirt-car is shown in Plate XII.; brick and cement were carried in on ordinary platform-cars.

The small lock extending through the roof of the caisson was designed for supplies. Its lower end was fitted with an extension which was just a quadrant of a circle; its cross-section was square, while that of the remaining portion was circular. One door was placed at the lower end of the lock, the other at the upper end about 3 feet above the ground. It was used principally for brick, the cement, plates, etc., being still taken in through the old horizontal lock. In loading it, the lower door having been closed, a barrel was filled with bricks and lowered by a rope and tackle; when at the head of the lock the bottom was dropped from the barrel, allowing its contents to fall out. The load varied from 50 to 100 barrels of bricks. After the upper door had been closed the lower one was opened, when the bricks fell upon the platform of the working-chamber. They were then taken to the heading upon a platform-car running upon a track extending down the north side of the tunnel—the track for the silt-car being upon the south side.

The economical and rapid removal of the excavated silt was a question of great importance. Previous to the accident a 6-inch blow-out pipe (Plate VI.) extended through the air-lock to the heading, the end being provided with a movable section and valve. Silt and water were mixed in a suitable receptacle into which the end of the pipe was dipped; upon the valve being opened the air-pressure forced the mixture out of the tunnel. The bottom of the shaft was first used as a receiver, but afterward a waste-tank upon the surface of the ground was designed. The silt was then lifted up the shaft in buckets and dumped into cars, which were drawn upon a narrow-gauge track running to the low land just back of the work, and there emptied. A covered tank was so constructed that the car could be run under a delivery-spout to be loaded, the tank being filled through the blow-out pipe.

As the tunnel was advanced it became necessary, owing to the increased friction in the pipe, to adopt some new plan for doing this work. The completed work was left a little more than one-half full of excavated silt, the intention being to remove this after the approaches had been opened, when it could be handled more easily, quickly, and cheaply; in the meantime it served, by its weight, to keep the new sections in position until the masonry had set. Upon the silt was laid a narrow-gauge track, upon which ran a dumping-car (Plate XIII.) having a capacity of 1½ cubic yards. The silt taken from the heading was then placed in this car, which was drawn up by a small engine located near the west wall of the working-chamber. At the upper end of the track there was a sharp incline of about 45 degrees, the rails on which were placed just wide enough to permit the front wheels of the car to pass between and remain on a horizontal track; the rear wheels, having faces two or three inches wider than those in front, ascended the incline, when the bottom of the car sloped sufficiently to allow its load to slide off into a well. (The operation of this car is shown very distinctly in Plate X.) The silt was then mixed with water and blown out by the air to the reservoir above ground.

An Eads sand-pump, the suction-pipe of which entered a box provided with a buffer, was introduced. The silt was spread over a wire screen at the bottom of the latter, and, being mixed with water, was discharged directly into the open air, thus doing away with the conveyance of the silt to the working-chamber on cars. The water and compressed air together disposed of the silt very effectually, and delivered it much faster than it could be supplied at the then (July, 1881) rate of excavation.

At one time, when the silt had been nearly blown out of the well at the working-chamber, and the end of the pipe almost uncovered, it was found impossible to turn the valve, some obstruction having clogged it. In the next instant the silt had all been driven out, and the air, making a terrific noise, rushed up the pipe. The men made haste for the air-lock, being led in their endeavors by one of the bosses. The superintendent, who chanced to be present, seized a shovel and placed it over the end of the pipe, against which it was tightly forced by the pressure of the air. As soon as the upward current had been stopped a piece of brick which had lodged in the valve dropped, when the escape was shut off. This incident is merely given to show how ready workmen engaged upon undertakings of this character are to take alarm. They knew they were in a perfectly safe chamber, and they also knew that it would take some time for all the air to escape and the water to reach them ; but ordinary workmen cannot be expected to stop and reason, especially when the retreat is ably led by one of their superiors.

The one who would attempt to direct operations in an undertaking of this kind must be gifted with both physical and moral courage ; he must not feel—for if he feels he will surely show—any timidity, and he must exercise such an influence over his subordinates that they will have confidence in him, in themselves, and in the success of his plans. He must have presence of mind ; he must quickly perceive and accurately realize the difficulty, and must unerringly judge of and adopt the best possible solution. While being willing to attempt extreme measures in extreme cases, he ought never to be willing to jeopardize his men or his work. Carefulness, thoughtfulness, and thoroughness must not be sacrificed for speed. While work was being carried on in the caisson Mr. Anderson, becoming tired out, stretched himself upon some cement-sacks and went to sleep. The work was uncertain, and the men were in such a condition as to become easily alarmed. In doing this he had two objects in view : he wanted to be instantly on hand in case anything happened, and he wanted particularly to inspire confidence in his men. When he was awakened he was surprised at the amount of work which had been done and the renewed cheerfulness of his assistants. One of them described the result to the writer as follows : "We thought it was all right when the boss went to sleep in there."

CHAPTER VI.

SHAPE, DIMENSIONS, AND USE OF PILOT—ADVANCING HEADING—OBJECTS OF IRON SHEATHING—MASONRY—LEAKS—BULKHEADS IN TUNNELS—CONNECTING WORKING-CHAMBER WITH SHAFT—PERFECTED METHOD OF OBTAINING SUPPLIES AND REMOVING SILT—SUSPENSION OF WORK ON NEW JERSEY SIDE.

In extending the work before the accident the outer sheathing of plates was put in position as soon as the excavation was made, and was in part held by struts resting upon the bottom and sides; this excluded the silt while the masonry was being built. Although the utmost celerity was practised in getting the plates and struts in position, the shell would at times be forced down and out of place sufficiently to make it almost impossible to maintain an accurate alignment, as we have already noticed. This was because the silt was not always of the same degree of compactness. At some points a material of a looser character was met, which, not having the sustaining power of pure, undisturbed silt, would exert a greater pressure upon the shell and bear it down. The expedient of inserting the plates higher than they should be was tried, but the variations in the weight of the upper material and its slight changes in density, together with the fluctuations of the resisting pressure, constituted data which were unreliable and from which no satisfactory results could be obtained. Nor was there any method by which the precise nature of every part of the surrounding envelope could be determined and the amount of its settlement gauged.

It therefore became necessary to introduce some device by which a firm foundation could be secured for the struts to rest upon. All the features which should be embodied in such a device seemed to be present in the "pilot,"* shown in place in the heading in the longitudinal elevation and cross-section, Plate XIV. This pilot was an iron tube, built up of plates, and of such a diameter as not to seriously occupy the space in the tunnel, and having a length sufficient to permit its forward end to enter some distance into the head-

* The "pilot" was invented by Mr. Anderson.

ing and its rear end to be abreast the completed tunnel or masonry; the intermediate points served as a centre from which to brace the plates.

It was composed of ten segmental, interchangeable plates 4 feet long, 22 inches wide, and ¼ inch thick, united by means of angle-irons riveted to the inner sides along the longitudinal edges; curved angle-irons were also riveted to the ends of each plate. In the joints were placed thick iron plates, which, projecting a few inches from the outer surface of the cylinder, formed series of transverse and longitudinal ribs which added greatly to its strength. Struts rested against the pilot next to the transverse plates, and supported the outer shell and masonry of the tunnel, and also the rear end of the pilot, which extended into the finished work. The plates were bolted together, and, as they were interchangeable, it was only necessary to have enough to make a tube of from 50 to 60 feet in length—25 or 30 feet being in advance of the heading and the remainder in the rear.

The longitudinal ribs stiffened the pilot, while the transverse ribs not only stiffened it but also acted as anchors to keep it in advance of the heading, for the enormous pressure tended to force it back into the completed section. When operating in material that will flow rapidly when mixed with a small quantity of water—as was the case with this silt—the tendency of the material is to run back along the smooth exterior of the pilot and thence down the heading. To obviate this the transverse joint-plates were adopted, and, as they could have been made of any width, it was always easy to obtain a surface of sufficient area to prevent any disposition of the earth to slide back along the cylinder.

By keeping the end of the pilot well in advance of the heading the ground through which the latter had to pass was very thoroughly explored, and when any soft spots or "pockets" were encountered they were much more easily and safely controlled than they could have been without the pilot. This was due to the fact that the exposed surface at the head of the pilot was but 6 feet in diameter, while that at the head of the tunnel was 21 by 23 feet.

In building tunnels through loose materials it has been customary to bank the head in order to form resting-places for the braces to support the crown of the arch. The immense pressure would

Plate X.

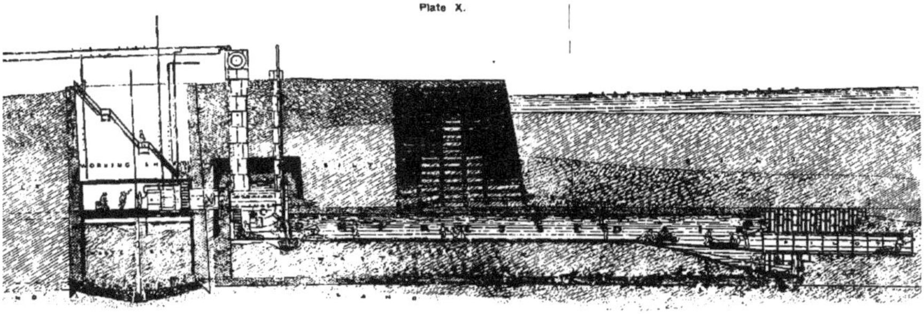

Plate XI.

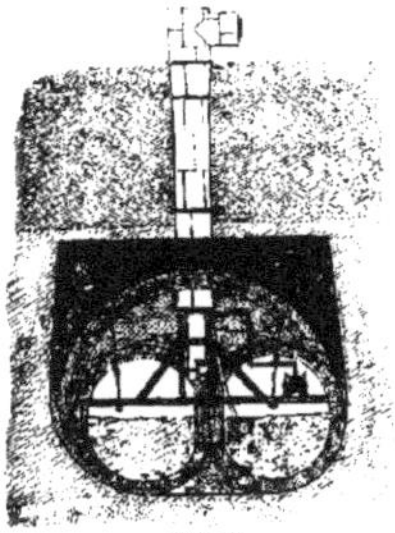

Plate XII.

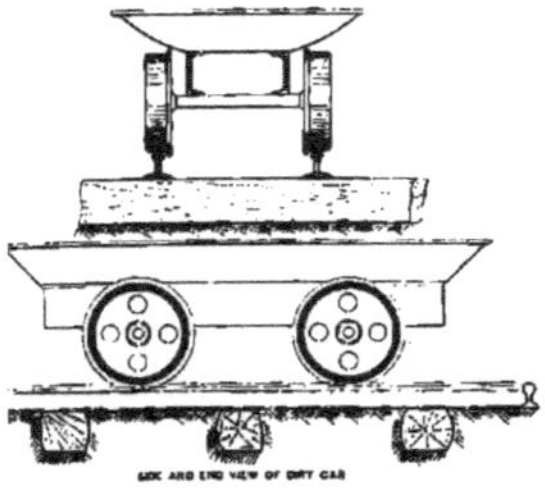

SIDE AND END VIEW OF DIRT CAR

Plate XIII.

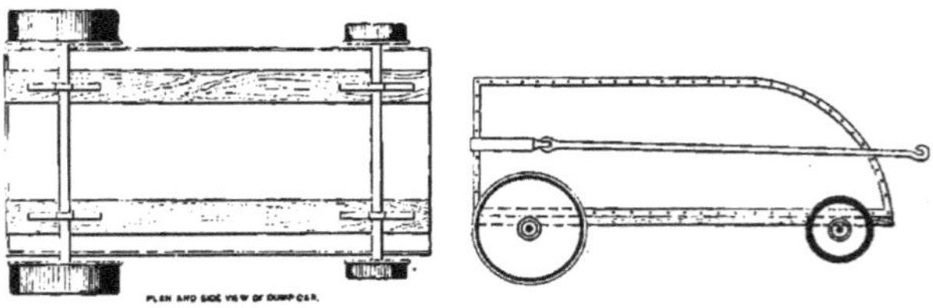

PLAN AND SIDE VIEW OF DUMP CAR.

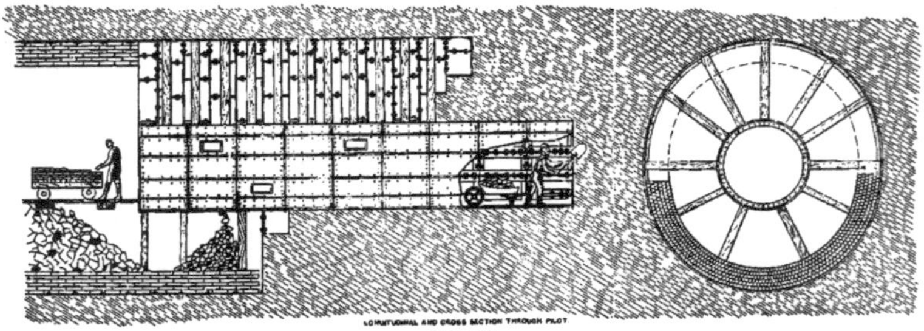

LONGITUDINAL AND CROSS SECTION THROUGH PILOT.

Plate XIV.

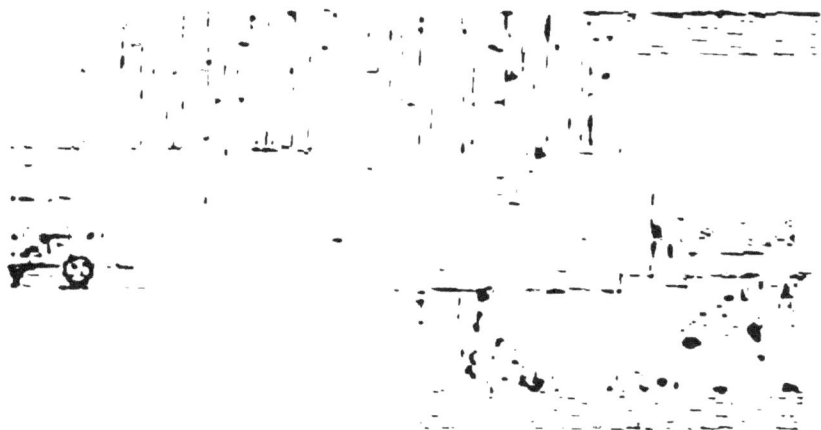

cause these to give way, and the large number of braces would so fill the working space as to interfere considerably with the men. The pilot absolutely prevented the earth at the heading from caving in, because the upper portion of the earth rested upon the pilot, and the space between the bottom of the pilot and the bottom of tunnel was not large enough to allow the earth to slide. As all the braces were radial ones, they were very much shorter than the radius of the tunnel, and, consequently, were of much smaller cross-section, and were more conveniently and rapidly handled than could have been the heavy braces used with tunnel-shields or with a banked heading.

There was ample room at the head of the pilot for two men to work without interfering with each other. In commencing a ring a space at the crown was dug out until a plate could be put in; this plate was bolted to the ring already completed. An excavation was then made at each side and the second and third plates inserted. Plates were taken from the rear and carried forward as they were needed. In this way the ring was built down each side and across the bottom. The transverse and longitudinal ribs were, of course, put in as the plates were raised. The excavated material was wheeled back and either thrown through side openings in the pilot into the finished section of masonry, or taken to the rear and there placed upon cars, to be removed in the usual way. The object in making holes in the pilot was to diminish the distance to be traveled by the men, thus lessening the time required to accomplished a certain distance. A two-board walk was laid from the inner end to the silt.

Actual experience demonstrated that the iron in the pilot was not of sufficient strength to resist the strain that came upon it, and the thickness of the plates was increased to ⅜ of an inch. The pilot was firmly supported at each end, but, owing mainly to its uncommon length, the central section was deflected when from any cause an excessive weight was brought to bear upon it; and although it sustained the radiating struts and received from them additional strength, this was not enough to insure the requisite rigidity throughout its entire length.

A very clear idea of the method of advancing the heading, as finally adopted, may be obtained from Plates X. and XIV. A space being opened at the top, a plate was put up, bolted to the one previously set, and supported by a brace resting on the pilot. As far as possible these braces were so located that their ends would be alongside of the transverse ribs and not in the middle of a pilot-plate.

Each ring was composed of fourteen plates ⅜ of an inch thick and 2½ feet wide. The six upper plates were 3 feet long and weighed about 170 pounds each, and the other plates were 6 feet long and weighed about 320 pounds each. These weights include the 3-inch angle-iron that was riveted to each edge with ½-inch rivets placed 6 inches between centres. The plates were fastened together, so as to break joints, with ¾-inch bolts placed 9 inches between centres.

This sheathing was not designed to act merely as a thin lining to prevent the passage of air or water, out or in; but, owing to the thickness of the plates, the evenly-distributed pressure of the air upon the interior, the pressure of the silt upon the exterior, and particularly to the stiffening effect of the angle-irons that were riveted to each edge, this coating had ample strength in itself to sustain great and unequal pressures upon the various parts of its surface. Although it was not intended to support, unaided, any force that might come upon it, it was intended to distribute the load evenly and to bear the greatest possible weight when assisted by the air-pressure and braces.

At first it was customary to dig out a space that would give plenty of room in which to insert the plate. Frequently an air-pocket would be left between the shell and silt, when the latter would begin a slow, regular movement toward the iron. Sometimes when the silt met the iron the latter would be under so great a load that it would be forced to yield more or less. To obviate this difficulty great care was exercised in completely filling all exterior holes, after the plates had been inserted and braced, with either silt or concrete, so that a settlement was prevented.

The excavated material was thrown into the last-completed section of tunnel, and when this had been a little more than half filled the rest of the material was loaded upon cars and hauled up to the working-chamber, where it was blown out, as previously described. To facilitate operations the track was extended upon an incline down to the lower part of the heading, which was cut so as to form terraces upon which the workmen could stand, each gang being particular to cut straight ahead in the wall and to preserve the silt platform upon which it operated.

As we have often remarked, the advance was made with extreme caution, to guard against the too sudden opening of a soft spot, where old piles had been withdrawn, containing water enough to enable it to flow as quickly as quicksand. A pocket was once opened in the heading near the arch of the shell. To prevent the inflow of silt a bulkhead was put up and braced from the masonry. A hole was made in the top of the bulkhead of a size to admit the hand, and a dry ball of silt was pushed in. A second was forced in, crowding the first backward and upward; continued repetition of this process formed a skin nearly impervious to air. This skin was then worked back with the hand, and as it was enlarged more balls were supplied, until the opening became of the required dimensions to admit a plate. During work of this kind considerable air would sometimes escape, making a very unpleasant noise and threatening a reduction in pressure and a consequent disturbance of the bulkhead, so that the operation required men of skill, courage, and coolness.

When the series of rings had been placed the shell was thoroughly cleaned of all dirt and the masons began the brick-work, which was rapidly carried up the sides and over the arch. The masonry was built in sections averaging about 12 feet in length. For the first 450 feet the two tunnels were built of hard-burned brick laid in the best Rosendale cement, in the proportion of one cement to one and one-half sand, the masonry being 2 feet thick. For a short distance the lower half of the tunnel was made of solid concrete, and the upper half was lined with a 5-inch course of asphalt blocks backed with brick-work, as an experiment; hard-burned brick was finally adopted, as being the best material with which to build the wall. When the north tunnel had been advanced for a distance of 700 feet from the shaft the thickness of the masonry was increased to 2½ feet, additional strength being deemed necessary because of the greater weight to which it would be submitted after the removal of the air-pressure; this, of course, being due to the increase in the depth of water.

For each foot of tunnel about 14 cubic yards of silt were excavated—one-half of this being taken out and disposed of as heretofore described, and the remainder being thrown back into the finished work. Each running foot of tunnel contained over 132 cubic feet of masonry, about 3,000 bricks, 6 barrels of cement, and 10 barrels of sand.

By the aid of the pilot the grade of the tunnel could be kept almost absolutely true, and any trifling variations that might occur could be readily rectified during the advance. It was preferably located at a point a little above the axis of the tunnel; and although this rule was by no means rigidly followed, experience proved that that was the best position to insure rapid work of good character prosecuted without risk. This left a large space under the pilot in which to move the excavated material from the heading, and did not take from its usefulness as a foundation for the braces, which, by this arrangement, could be more easily put in, since the shortest ones supported the roof-plates and the longest ones the plates for the invert.

During all these operations, in fact at every stage of the work, a constant vigilance was enforced to detect and stop leaks. Air escaping through a minute hole made no audible sound, but the opening, unless soon stopped, grew larger and larger with great rapidity. The flow through a single hole was insignificant, but when these holes were closely distributed over a large surface the quantity of air lost became of importance. Yet this care was not exercised with a view toward saving the air thus lost. The atmosphere in a chamber such as was formed at the heading, and in which many men were working, soon became impure unless it could be often replaced by fresh air; the only passages through which the vitiated air could pass out were the leaks and through the lock placed in a bulkhead in the tunnel, as we shall presently describe. The principal object, and one on which the temporary welfare of the work depended, was to prevent the escape-openings from assuming such proportions as to be beyond control, since in that case a blow-out would be the result and the tunnel would be flooded. This inspection extended from the uncovered silt at the heading back to the masonry, and embraced every spot at which escape of air was possible. Although men were appointed to perform this labor, it was the duty of each one to be on the look-out.

A bulkhead was built in both the north and south tunnels at a distance of 430 feet from the shaft. This step was necessary, since the leakage through the long line of brick-work had become troublesome, and the air-pressure needed to keep the silt in place at the deepest points at the headings too great to be safely applied to the tunnels at their highest parts. This formed two independent air-chambers at the headings of the tunnels. The bulkheads were brick walls, 4 feet thick, let into the tunnels, as shown in the longitudinal and cross-sectional drawing, Plate XV. Each wall was backed by a solid wall of horizontally-placed timbers 12 inches square, against which rested vertical timbers, which were strongly secured by braces let into the permanent masonry. Two air-locks, 15 feet long by 6 feet in diameter, and similar in general plan to the first air-lock in the shaft, were put in each bulkhead. The construction of the air-lock is very clearly shown in Plate XVI. The plates were bolted together in order that the lock could be taken down, transferred to a more advanced bulkhead, and re-assembled. It was designed that one of these locks should always be open toward the heading, in order to afford a place of refuge for the men in case of necessity; the second lock was used for the passage of supplies, etc.

After the completion of these bulkheads the tunnels between them and the shaft were thrown open to the air. It may be well to state here that this portion of the work has been open for very nearly three years, and has shown no signs of displacement, and the masonry is as perfect as when first built.

No attempt was made to connect the working-chamber with the shaft—cutting through the west side of the caisson—until the bulkheads had been put in; but the inconvenience resulting from the necessity of taking all the supplies through the caisson-locks compelled the finishing of this section as soon as possible. While this task presented none of the obstacles that had long been confidently anticipated, it was sufficiently troublesome to keep up a lively interest in the minds not only of those engaged in it, but also of all those at work in the east heading of the north tunnel, 700 feet distant.

After discussing various plans, one proposed by Mr. S. H. Finch, assistant engineer, was adopted. This plan was simply to close the doors of the old air-lock and shafts leading to the tunnel, admit the compressed air, and prosecute the work in the same way as at the heading. Preparations were made by sending down several days' supply of brick, sand, broken stone, and cement, after which the doors were shut and the pressure raised to 9 pounds per square inch. The lowest section of the material through which the work had to pass was sand, the middle section was silt, and the upper silt and cinders. These had been undisturbed for several months, and, it was thought, had become compact enough to permit work. It will be remembered that the caisson was about 3 feet less in length than the shortest distance between the shaft and tunnels, and at the sides of the shaft about 15 feet less, so that the connecting tunnels were wedge-shape in plan, one side being curved to conform to the contour of the shaft.

The false brick-work which had been put in to enclose the caisson was taken out, as were also the timbers on that side of the caisson. The material was found to be quite dry and very compact, although it was a mixture of ashes and silt in about equal parts. No trouble was experienced in placing the roof-plates, and the work proceeded rapidly, but with extreme caution. After the excavation had been made, the iron shell and a portion of the invert of the south connections put in, and when the silt had been removed in the north connection to within about 3 feet of the bottom, the water came in so quickly as to compel a cessation of work. A 2-inch blow-out being of no use, the pump in the shaft was connected with the stand-pipe and the full head of steam turned on, while the air-pressure was increased to 11 pounds per square inch, when the water suddenly disappeared. The remaining silt was then removed, but before the plates could be connected at the bottom the water again rushed in, but receded with the tide, leaving the bottom dry, when the masonry was laid without further trouble. Afterward the false piece in the shaft was taken out, thereby completing the work. Both the filling and sand through which the connecting tunnels were carried contained abundance of water, and, the east heading being the lower, it was feared, in the event of water breaking in freely, that the air-locks leading to that heading might become submerged, and thus close the only avenue of escape of some forty men who were engaged in pushing forward toward the middle of the river.

Plate XV.

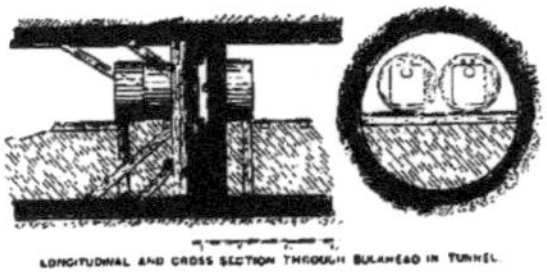

LONGITUDINAL AND CROSS SECTION THROUGH BULKHEAD IN TUNNEL

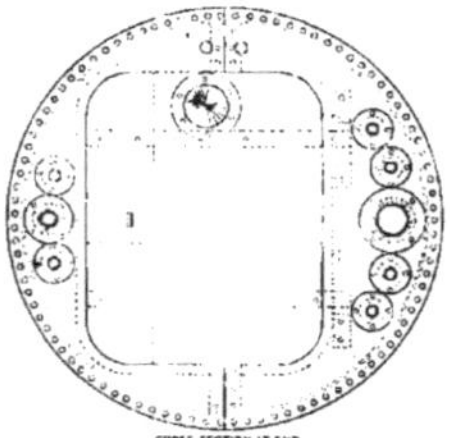

CROSS SECTION AT END

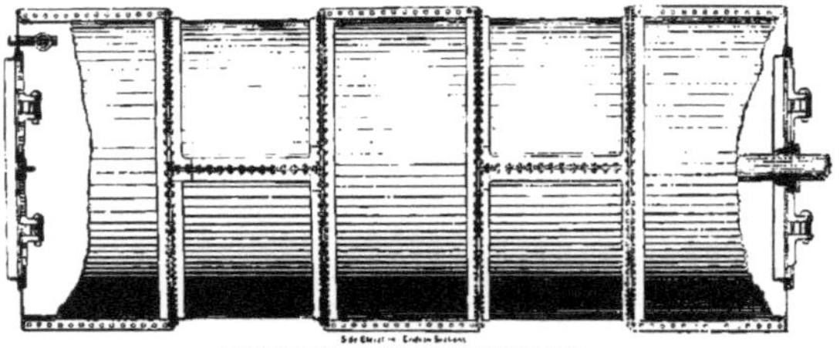

Side Elevation Ends in Sections

HORIZONTAL AIR LOCK - MADE IN SECTIONS TO FACILITATE MOVING TO FRONT

Plate XVI.

Plate XVII.

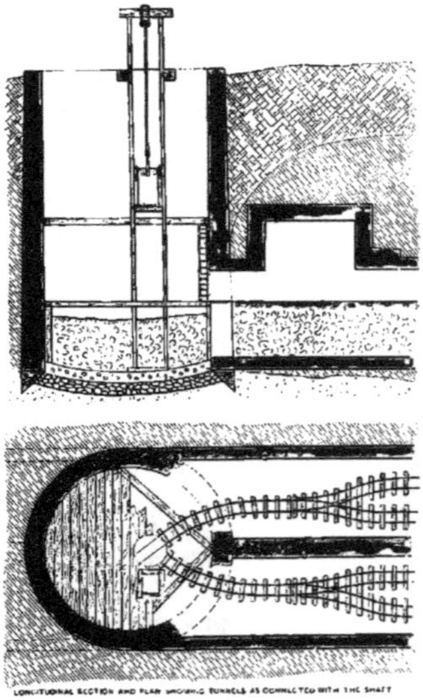

LONGITUDINAL SECTION AND PLAN SHOWING TUNNELS AS CONNECTED WITH THE SHAFT

Plate XV.

Plate XVI.

During the entire time those men were not interrupted for one moment. Plate XVII. shows the tunnels as finally and permanently connected with the shaft.

A platform had been built in the shaft at a height of about two-thirds that of the tunnels; this was the general distributing point for material passing either way. A single track led from the shaft to each tunnel, where it met the double track extending down to the locks. An elevator was located just east of the centre of the shaft; cars loaded with supplies were run upon the cage, which was then lowered to the platform, where they were run upon the track leading to the lock in the bulkhead, through which they were passed to the heading. Each car was provided with a brake with which to govern its speed, the descent being made by gravity. As the cars were brought up the tunnel they were run upon the elevator, raised to the surface, and run upon a track leading to the low lands; by this plan the silt was handled but once, and much time and labor were saved.

In the north tunnel, at a distance of 280 feet from the first bulkhead, a second one was built similar to the first, with the exception that it contained but one air-lock, which was deemed sufficient, since the air-pressure in the chamber between the two bulkheads was kept at but a few pounds—an average of about 6—less than that required at the heading.

Material was drawn from the heading to the second lock and thence to the first by a horse and a mule. These animals had with great difficulty been taken through the air-lock, and had remained under pressure varying from 20 to 33 pounds per square inch for several months without showing any signs of suffering; in fact, the horse when introduced was almost useless, since he was old and severely afflicted with the heaves, but while in the tunnel he improved much, having been cured of the heaves and having gained in flesh, although almost continually at work. Upon their removal, which was made necessary by the cessation of operations, the mule, which had been under pressure about three months, was liberally dosed with ergot, whiskey and ginger, and in a few hours was as well as ever. But not so with the old horse, that had been in seven months. He was subjected to the same treatment, but the change had been made too suddenly (he was taken out in about half an hour), for in a few minutes he was dead.

In November, 1882, work on the New Jersey side was stopped because of lack of funds. This was greatly to be regretted, as the work was progressing very rapidly and to the complete satisfaction of all interested. During the last four weeks 127 feet of finished tunnel had been put in; this was an average of 4½ feet each day. Had operations continued, in all probability the north tunnel would have been completed six months since. The north tunnel had been built for a distance of 1,550 feet from the shaft and the south tunnel for 570 feet. The headings were securely bulkheaded, and for some time the finished sections were kept free of water; but, as the exact date of resumption could not be determined, it was thought best to devote all the energies to pushing forward the work from the New York side. Consequently the pumps were stopped, when the water very slowly found its way into the tunnels from leaks around the shaft, flooding the whole in about three months; this is the condition of the undertaking at the present writing (December, 1884).

Before the water was permitted an entrance the bulkheads in the tunnels were so firmly strengthened that their giving away, or becoming weakened from any cause, was rendered impossible. The pumping machinery is so disposed that it can be set in operation at a moment's notice, and it is powerful enough to relieve the tunnel of water in two or three days.

CHAPTER VII.

SILT, ANALYSIS AND PHYSICAL PROPERTIES—ITS RESISTANCE TO DISPLACEMENT—EFFECT OF COMPRESSED-AIR UPON SILT—DRY, MOIST, AND SATURATED SILT—THE USE OF COMPRESSED-AIR—STOPPING LEAKS—EFFECT OF COMPRESSED-AIR UPON THE MEN, AND SENSATIONS IN PASSING THROUGH THE AIR-LOCK.

It is extremely doubtful if a substance could be found which would be better adapted to submarine tunneling by compressed air than the so-called silt that constitutes the bed of the Hudson River. As there found it is of very uniform character, extends entirely across the river, and is of sufficient depth to wholly surround the tunnel, except for a distance of about 900 feet on the New York side, where, in places, it will cover but about two-thirds of the structure, leaving from 2 to 7 feet of the lower side embedded in sand or rock.

It is a washing brought by the water from the upper regions drained by the river, and the magnitude of the deposit, in the area covered and depth (varying from 28 to 100 feet), indicates that the work has been going on for ages.

An analysis made by A. R. Leeds, Professor of Chemistry at Stevens Institute, gave the following result:

	Per cent.
Combined water,	5.13
" silica,	58.95
Free silica, or quartz,	10.32
Alumina,	15.14
Protoxide of manganese,	0.95
" iron,	3.28
Sesquioxide of iron,	1.38
Lime,	2.88
Magnesia,	1.50
Sodium combined as chloride,	0.23
Chlorine existing in form of chloride,	0.38
Sulphuric acid,	trace.
Titanic "	trace.

Silt is of a dark slate-color, has very slight cohesive properties when dry—it may be easily crumbled in the hand—is in the form of an impalpable powder, is very sparingly mixed with hard, rocky substances which, when they do occur, bear a close resemblance to, and in some cases are, petrifactions of wood. Shells of a small size are not infrequently found in it, but gravel-stones and small boulders are seldom encountered. When it carries the proper degree of moisture it forms a compact, dense, tenacious mass, having a most effective power of cohesion that enables it to retain a given shape for an extended period of time; in this state it may be handled much as ordinary putty can, and which it very closely resembles both in feeling and in the after-effect produced upon the hands. It is not dirty, in the usual sense of this word; a little water quickly relieves the hands of its presence, and the flesh has a soft, almost oily, feeling afterward. The most important feature in its relation to the work we are describing, and one directly the outcome of its cohesiveness when of the proper degree of saturation, is that both air and water pass through it very slowly. When well mixed with a sufficient volume of water it flows as freely as quicksand and is much more difficult to control.

In a paper read by General William Sooy Smith before the Western Society of Engineers we find the following paragraph: "When exposed to air-pressure of 20 pounds to the square inch it requires an additional pressure of 2,700 pounds to the square foot to force a disk into it, this pressure being accompanied by vibration. Its total resistance to displacement at a depth of 60 feet below water-surface (at which depth experiments were made to determine this resistance) would therefore seem to be 5,580 pounds per square foot. This is relied upon to insure permanency of the work under the conditions of actual use. It is easy to make a cut into it 2½ by 5 feet, and 5 feet deep, in order to place in position one of the plates of the iron shell."

It was impossible to force the hand, either clenched or open, to any considerable depth into the silt at the exposed heading of the tunnel; and when the surface was struck a heavy blow with the fist a dull sound was produced, and hardly any depression was formed beyond the mere imprint. The steps into which the heading was cut retained their sharp outlines amply long enough to enable the men to place the plates in the upper portion of a ring and excavate a new series of steps. It is impossible for moist silt, because of its very nature, to permit an air-pocket to remain in it. This was clearly indicated by the settling which took place when spaces were left between the crown-plates and the undisturbed silt, and when the south tunnel upon the New Jersey side was started by being built around a core of silt. The settlement took place gradually, slowly, evenly and surely; the rate of any settling movement always seemed to depend directly upon the amount of water distributed in the silt through mechanical disturbance—when the water was largely in excess the silt flowed in a stream; when comparatively dry it crumbled, and large pieces would fall.

Compressed air affected the silt by driving the water back in it; the distance to which the water was forced, and what we might term the thoroughness with which this operation was performed, depended upon the ratio existing between the amount of air-pressure and the hydrostatic head. As we have previously mentioned, it was not essential to maintain an air-pressure in the work equal to, or in excess of, the water-pressure in order to accomplish the best results; the air-pressure was almost invariably kept a little less than the water-pressure. This slight difference, which was seemingly reversed, varied according to the density, or compactness, of the silt worked through, and consequently it fluctuated from one extreme to the other. Of course the distance to which the air penetrated in the silt, pushing the water before it, was influenced by the amount of pressure—an air-pressure but one pound lower than the water-pressure forced the water out more rapidly and dried the silt to a greater depth than an air-pressure of 5 pounds less. Therefore the only method, having absolute reliability, of determining the proper air-pressure was to vigilantly and continually watch the exposed silt in the heading. A smooth, even, and, in a certain way, a glossy surface, when rubbed with a shovel or with the hand, indicated that the silt and water were mingled in correct proportion. A tendency toward hurtful dryness was first shown by the appearance of small, pin-like apertures through which the air might, or might not, be escaping; this was remedied by decreasing the air-pressure. A wet, clammy surface, down which the water trickled, showed that the water was in the ascendency and should be forced back by an increased air-pressure. These might be said to embrace the scale of variations beyond which it was dangerous to go either way; the drying process might continue until a great mass of silt became detached, thereby opening so large a passage for the air to escape through as to be beyond control; or the silt might become so thoroughly saturated with water as to flow in in a large stream. The air within the tunnel was very moist, and the warmth—the temperature was about 75 degrees—caused a condensation upon the cold surface—normally about 55 degrees—of the freshly-exposed silt, which was often mistaken for water coming in through the silt; a little excess of pressure would force this water back and thus cause a dry surface.

A most important fact may be here noted: to effect these changes required considerable time, and by the exercise of ordinary caution could always be detected in season to prevent mischief. It was not possible to instantly dry or instantly saturate the silt. This feature guaranteed the safety of the men and work, unless the former, owing to long-continued immunity from anything even remotely resembling a blow-out, should become careless and neglect measures tending to their own welfare.

Silt, as found in its natural position, was practically impervious to both air and water. This feature was clearly shown when it took the air four days to pass through 8 feet of silt next to the shaft, the air-pressure being 2 pounds greater than the hydrostatic head. A cup made of silt, one inch thick, would hold water for several days if not disturbed, but a very slight jar would greatly change its character.

The air-pressure served as a support for the plates which not only acted as an additional guard to confine the air, but which possessed, when in the form of a completed ring, great strength to resist external strain; they also served to distribute the load more evenly. The four angle-irons riveted to each plate, and the method of uniting them, increased their capability to withstand bending, and most materially stiffened the shell as a whole. In the sand upon the New York side the plates alone confined the air, the sand offering no resistance to its passage. If the shell, then, had been dependent solely upon braces for support—the balancing air-pressure all being removed—it would have been absolutely impossible to have made any advance whatever.

Small, incipient leaks in silt were stopped by application of wet silt; in openings demanding stronger treatment bags of cement were sometimes inserted. The air rushing through a hole would catch and hold the material placed in that hole. (The writer is well aware that this feature of tunneling by compressed air has been much disputed, but it is nevertheless true, and it has been done so many times that it cannot be controverted.) A disk of paper or a piece of leather held at the hole would be sustained by the air. The temporary closing of the opening was oftentimes sufficient to prevent further escape. The explanation is simple: as soon as the opening had been closed, the silt, through which the air had forced and kept open a channel, settled back to its original position, and the compactness of this portion depended upon the time during which the silt was allowed to settle. This was illustrated at one time very forcibly by Mr. Anderson, while working under the dock at the New Jersey side, thrusting his shoulders into a hole which was quickly assuming dangerous proportions. By working his shoulders and forcing them up he was soon enabled to "fit" himself into the aperture—the instant at which the fit was accomplished being indicated by the pressure catching and holding him. The silt fell back and in a short time was dense enough to admit of the usual method—forcing in silt—being carried out.

That the men stood the compressed air so well was in a great measure due to the admirable precautions and regulations of the company. Each man was examined by a physician before he was allowed to enter the work, and all not in robust health, particularly those suffering from lung or heart trouble, were rejected. The first time a man passed through the lock the pressure was admitted very slowly and the effect upon him carefully noted. He was told how to relieve the unpleasant ear effect, and if there was no further trouble he was introduced to the heading. A person being subjected to gradually increasing air-pressure experiences a hard pressure upon the interior of the ears, accompanied by more or less pain. Coughing, violently blowing the nose, swallowing, etc., will relieve it for the time being, but as the pressure increases it returns, when swallowing is again resorted to. As the men emerged from the tunnel all violent exercise was forbidden; but as some of the stronger ones would run up the stairs in the shaft (New

Jersey side), a rule was laid down that all should be taken up on the elevator. Upon the New York side, the air-lock being at the top of the shaft and near the surface, no climbing was necessary after coming out.

The effect produced by entering compressed air is to start a profuse perspiration, to increase the heart-beats but make them weaker, and to exhilarate the individual. The sense of smell is so weakened as to only detect the most powerful odors, or is completely annulled; the other senses seem to be unaffected. Upon leaving compressed air a person feels weakened for a time and is averse to exertion; this soon passes away. Men laboring in an air-pressure need more sleep and sleep sounder than those working in the open air. Compressed air should neither be entered with a full nor an empty stomach; an hour or two after the meal is the best time. The use of liquors was prohibited (unfortunately there was no way of enforcing this rule); strong coffee was given the men as they came from the lock.

There were comparatively few cases of caisson disease, none of which was fatal. The suffering was mainly confined to the limbs, the body being but infrequently affected. Upon the New Jersey side the use of two air-locks, placed two or three hundred feet apart, lessened the danger of being afflicted with this disease, sometimes designated the "bends." After passing through the first lock a person was under a moderate pressure, which was kept at about two-thirds that in the heading. The walk from this lock to the second accustomed him to the change, and he was less affected when passing through the second. The highest pressure used was 34½ pounds per square inch (this was exceptional) at the heading, that between the locks being kept at about 25 pounds. That continued exposure to compressed air is not injurious is shown by the horse and mule heretofore mentioned, which not only lived for months in the tunnel, but kept in good condition and worked hard almost constantly day and night.

CHAPTER VIII.

PLANS FOR REACHING GRADE OF TUNNELS AT NEW YORK SIDE—SHAFT OR CAISSON—DESCRIPTION OF CAISSON—AIR-LOCKS—ITS LOCATION—METHOD OF SINKING—PLANS FOR STARTING TUNNELS—CUTTING THROUGH SIDE OF CAISSON—IRON BULKHEAD—SIZES OF PLATES—METHOD OF BUILDING THE TUNNEL—BULKHEAD IN TUNNEL—HANDLING MATERIAL—SAND-PUMP—BLOW-OUT—PILES MET—WORK STOPPED—BOILERS AND AIR-COMPRESSORS—MATERIAL REQUIRED TO COMPLETE TUNNELS.

IN this, as in other great works characterized by the marked originality of the plans pursued, experience has been the best teacher and most trustworthy guide. When selecting methods for accomplishing a certain object no set rule applicable in all cases could be deduced. The governing conditions changed continually, and it was impossible to calculate the direction of those changes; the time at which features compelling modifications in the plans would present themselves was always an unknown factor. We find that, before the exact nature of the material constituting the bed of the river had been ascertained, the shaft seemed to offer by far the cheapest, quickest and safest means of reaching the requisite depth at which to start the tunnels; and, as this had worked well on the New Jersey side, plans were made for a brick shaft by the aid of which to reach the grade of the tunnels at the New York end. This was to have been similar in form to the first one and 42 feet inside diameter, but in addition it was to have had an iron shoe and shell outside extending up 35 feet from the bottom, and so arranged that the lower 25 feet could be converted into a caisson by putting on a diaphragm or roof. It was designed to remove this roof after the tunnels had been started and the bulkheads and air-locks put in.

There was much discussion whether to adopt this plan or simply a caisson built of wood, and, as usual in such cases, each plan had many advocates. The timber caisson was finally decided upon as being the better, although experience had shown that it caused much inconvenience in getting material out and in. It was also arranged to cut a shaft through the roof of the caisson as soon as the tunnels had been extended far enough to permit of the placing of the air-locks; this would convert the caisson into a spacious and convenient working-chamber. There can be no question as to the advantages to be derived from a large, roomy working-chamber at the entrance to the tunnels to serve as a centre from which to distribute supplies and material.

Work upon the caisson shown in Plate XIX. was begun in July, 1881. The bottom was 48 feet long and 29½ feet wide; it was 25 feet high, and the sides had a batter of one-half to one and were 3 feet thick. The top was 46 feet long and 27½ feet wide. The transverse struts were 16 by 18 inches; the longitudinal ones were 20 by 20 inches; the vertical ones 16 by 18 inches. Transverse and longitudinal tie-rods provided with turn-buckles extended through the structure from outside to outside. The lower edges of the sides were bevelled, and shod with iron to form a cutting surface that would resist wear.

It will be seen that this caisson resembled the one used at the other side of the river only in the form of its exterior. It was a much more difficult structure to make secure—the heavy timber arch in the other, together with the triangular sections that were filled with concrete, greatly increased the strength—and the shape of its interior was not so well adapted to subsequent operations. In addition to this it cost much more to build.

Practice upon the other side had suggested many improvements in the number and disposition of the air-locks, and extending through the roof of the caisson were three shafts provided with air-locks. In the centre of the roof was a shaft 5 feet in diameter, to the lower end of which was attached a cross-piece 6 feet in diameter and 15 feet long. At each end of the cross-piece was a door 3 feet wide by 4 feet high, and at the lower end of the shaft was a third door. Of course all these doors opened toward the interior, or air-pressure, and when closed the joints were made air-tight by means of rubber packing. This shaft extended but a few feet above the caisson; a wooden box slightly flaring toward the top extended to the surface of the ground. At one side of this shaft was a second one, 3 feet in diameter and extending just through the roof, and furnished with a door at each end. Upon the other side was a third shaft, 5 feet in diameter, at the upper end of which —about 15 feet below the surface—was the air-lock used upon the New Jersey caisson, which we have already described, and which is illustrated in Plate IX. The central air-lock was used for the passage of general supplies and excavated material; the smallest shaft, built long and narrow, was used in getting in timbers, which it would admit in long lengths: the third air-lock was for the men.

The caisson was located so that the side next the river was 60 feet from the bulkhead-wall. It will be remembered that the descent of the caisson upon the New Jersey side was guided and controlled by six suspender-rods, upon which screwed nuts resting upon heavy timbers projecting over the edge of the coffer-dam and extending back upon the ground, the rear ends being held down by heavy weights. This arrangement was considered superfluous upon the New York side, and no device was used to control or guide the caisson as it descended. The sides were extended upward to form a box or coffer-dam which contained the enormous weight necessary to force the structure down. The weight of load was 2,100 tons and of caisson 400 tons, a total of 2,500 tons. A pressure of air in the interior prevented the water from rising above the line of the shoes or lower edge; of course this pressure was increased as the depth became greater.

The earth was carefully removed from under the shoes, and the air-pressure—which practically formed a cushion upon which the caisson rested—lowered, when the weight overcame the friction of the earth upon the sides and the entire structure dropped. These advances were made through only short distances at a time, and great care was exercised in having the trench under the shoe perfectly level and as free as possible from large stones, in order to prevent all liability of straining the caisson by unequal sinking. After having passed through loose filling and dock-mud the lower edge rested in sand at a depth of 62 feet from the ground-surface and 3 feet above the exterior of the invert of the tunnels. The top of the coffer-dam was cleaned out to a depth of about 15 feet, and the sides were firmly secured by cross-braces, and a platform built.

Through the centre of this chamber passed the wooden shaft leading to the supply-lock. The timber-lock reached nearly to the platform, and the air-lock for the men was just above the platform, from which a ladder led to the surface.

Almost the whole of the caisson was in sand, and the two tunnels would be entirely embedded in sand and gravel. It will be seen that the material was very different from that encountered at the other side of the river, and instead of facilitating the work it proved to be—as it generally is in undertakings of this kind—one of the most difficult to handle. Some parts of the following paragraph, published * by the writer after both tunnels had been carried a short distance from the caisson, are applicable to many stages of this work, which has been most freely criticised and most sparingly studied: "The fact that the caisson was embedded in sand, led to the belief among many engineers of high standing that an outlet could not be obtained and the tunnel started by the system of working by compressed air. Indeed, it has become unsafe to pronounce an unfavorable opinion in regard to any particular piece of work connected with the tunnel; in more than one instance obstacles which seemed to present an insurmountable barrier to all future progress have been met, conquered, and the work has gone forward. New devices and plans have kept pace with new difficulties. At a first glance the sand above mentioned seemed to contain all the characteristics requisite for a first-class insurmountable obstacle. Upon the least reduction of the air-pressure this material would follow the water into the caisson; the smallest opening afforded a ready passage. The water and sand could be kept quiet as far down as the air-pressure was carried, and no further; and if a trench were dug or undertaken the upward pressure, due to the difference between the air-pressure and head of water, or depth of excavation, would fill the trench with sand and water about as fast as it could be taken out, and the adjacent material would then be in no better condition than at first."

The iron plates upon which the masonry was to rest were laid along the north and south sides of the caisson, being carried down in horizontal rows on the curve desired for the invert. Each plate was braced as it was put in, and the pressure was kept a little below

* *Engineering News*, Sept. 2, 1882.

the hydrostatic head, or about 25 pounds per square inch. Here again the system of iron plates with angles attached was found to be of very great value. After this iron shell had been finished the brick-work was laid. Thus was provided a working-chamber, or base of operations, similar in many respects to the one at the other shore. The work was relieved of water by means of a rotary pump located at the surface and connected with a stand-pipe leading into the caisson, where it joined three pipes extending to different parts of the chamber.

To cut through the side of the caisson and build the first section of tunnel was now the difficult task. This was by far the most formidable obstacle which the engineers had had to surmount. The material being sand and gravel, it was known that the method used so successfully when working through pure silt would be of no service here unless radically altered. It was determined to use compressed air to counterbalance the exterior pressure of sand and water; but as the sand possessed no self-sustaining qualities—it was as unstable, practically, as the water surrounding it—it became essential to interpose a sheathing between the air-chamber and sand. While inserting the separate sections composing this lining, whether they were of wood or iron plates, it would be of vital importance to hold back the sand temporarily; and since the joints could not be made perfectly impervious to water without delaying the work too much, some plan would have to be devised for removing that which would come in. The stand-pipes above referred to were expected to do this.

One plan which was believed would give the best results was to construct a movable bulkhead of oak plank, hold it at the ends by heading-props and braces from the side of the caisson or face of the completed work, as the case might be, and extend it from about 6 inches above the invert to the roof or crown-plates. The opening at the bottom was to permit the water to flow from the heading and be pumped out. It was expected that considerable would drain away, and that the surroundings would thereby be improved. Advance was to be made in this way: The air-pressure being maintained at about 1½ pounds per square inch more than what would be required to keep out water at the crown, the upper plank would be removed, enough material taken out to admit a plate, when the plank would be moved forward and supported against the next set of supports. The adjoining plank would then be advanced in the same manner, and so on down to the bottom.

In actual practice iron plates and angles were substituted for plank, and the conclusion was afterward arrived at that no progress could have been made if this had not been done, from the mere fact that a bulkhead of iron could be made almost perfectly air-tight—a feature not possible with one of wood.

The actual work of cutting through the side of the caisson may be described as follows: A score, or mark (shown in the dotted lines in the longitudinal section, Plate XIX.), had been made on the inner side of the caisson, of the exact form and size of the tunnel, at the time of its construction, and no spikes had been driven within 6 inches of this mark. After the invert had been completed men were set at work boring with 2 and 2½-inch augers on that mark, the air-pressure being kept at about 18 pounds. As soon as the calking course of 4-inch planking had been pierced the air escaped in small quantities; but after the augers had reached the diagonal planking between the 12 by 12-inch timbers the air rushed out with a loud noise. This was stopped by blowing dry cement into the holes whenever the leak occurred. This blowing consisted in holding a lot of loose cement near the aperture, into which the escaping air would carry it. Usually two or three quarts were sufficient, but in the case of the first openings as many barrels were used. Wooden plugs were also provided, and a number inserted in the holes immediately after the removal of the augers; but the cement proved convenient and was generally used. The boring was continued until nearly the entire circumference had been penetrated, when the work of cutting out was begun. Owing to the number and arrangement of the spikes, dowels, tenons, etc., employed to fasten each piece of timber, this was a labor of great magnitude, requiring the exercise of patience and perseverance. The caisson wall was 3 feet thick, of yellow pine, and as solid as it was possible to make it. As soon as a hole had been made large enough to admit it a plate was inserted and held firmly in place by braces. To keep the exposed portion from flaking and falling before the plate could be placed in position planks were held against it.

At a distance of 12 feet from the side of the caisson a bulkhead

Plate XIX.

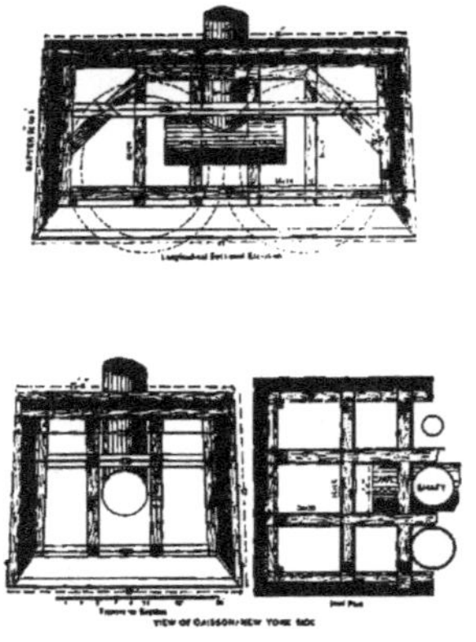

Plate XX.

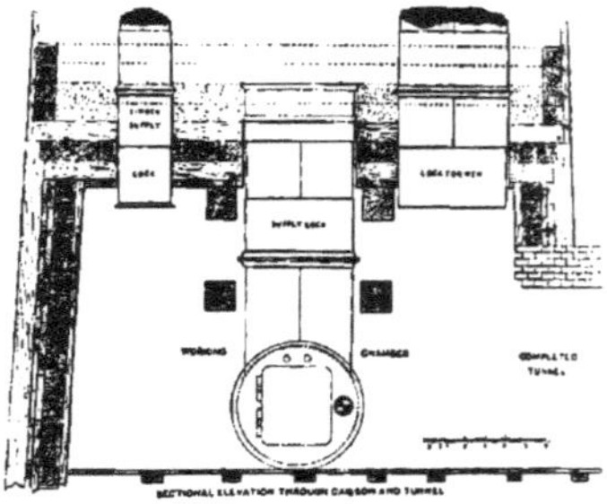

Plate XVIII

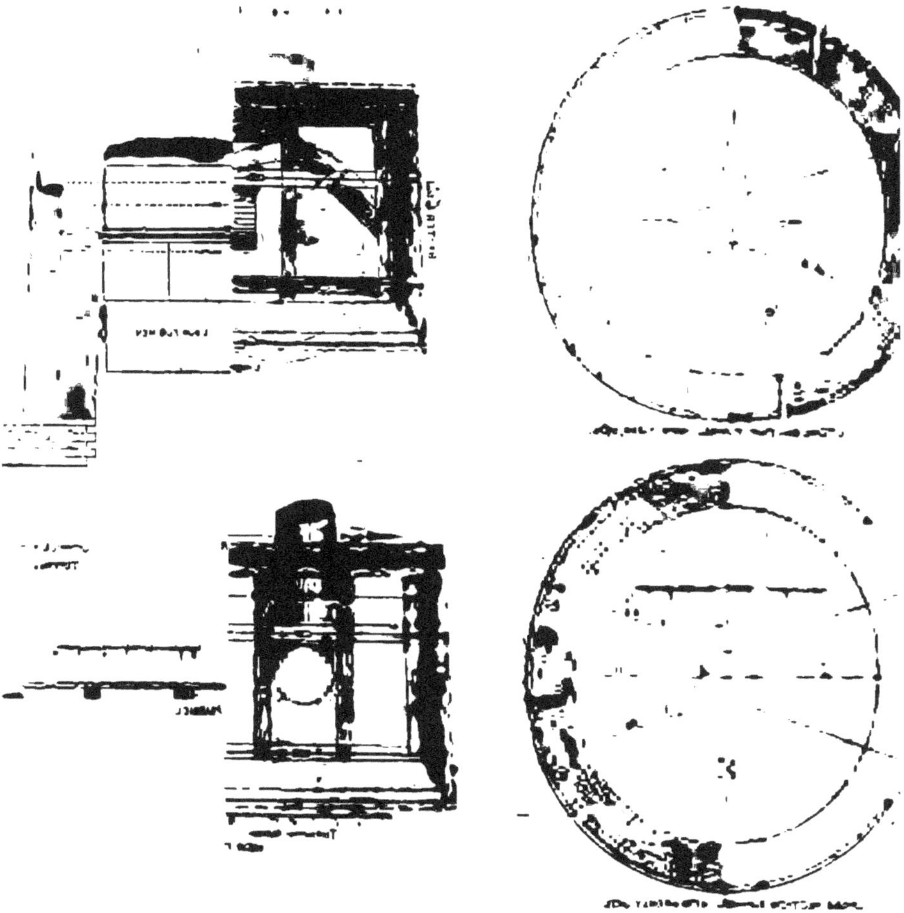

of iron plates was started at the top and built down. A system of struts, resting against the caisson, supported this bulkhead, which was put up in regular courses of 15 inches each, the plates being 4 feet in length, except those for the ends of each course, which were cut to correspond with the section of tunnel at the respective courses. It was found impossible to advance a ring of plates of the usual width—2½ feet—so they were changed to one-half this width, or 15 inches. This was the first modification or change worthy of note from the practice on the New Jersey side, and, although it necessitated more labor, it was beneficial because of the increased number of angle-bars, in one direction, which nearly doubled the strength. The proper thickness of the plates was a subject much discussed at the time, it being maintained by some of those connected with the work that it could be done without great or unusual risk with plates of any thickness (those used were ¼ of an inch), and that those in use were sufficient, and, being light and easily handled, it was expected to hold them in place by the exercise of skill and judgment in bracing. Before the heading had been far advanced it was found expedient to further reduce the size of some of the plates, for reasons explained a little further on.

The method of building the tunnel may be briefly summed up as follows: A chamber, generally 10 feet long and in cross-section equal to the exterior of the tunnel, 23 feet, was lined with iron plates which had been inserted one at a time, beginning at the crown of the arch and extending down the sides and heading; the rear end of this chamber opened into the completed work. After having been thoroughly cleaned the brick-work was laid. The amount of care, skill, patience, and perseverance which were necessarily exercised in prosecuting this section of the work successfully will be fully comprehended by all those who have ever had their undertakings hazarded by quicksand 65 feet under water.

The operation of advancing the plates to form the shell was an easy matter, so far as the roof-plates were concerned—about one-fourth of the circumference—although great care and skill were required. The plan was to drive light "poling" strips, of ½ by 2 inches and about 2 feet long, directly in front of the last plate put in; the rear ends of the strips were pushed up above and then pulled back so as to rest on the plate already in place. This prevented the material from falling while the excavation was being made; where the air would escape too much a plaster of silt, put on over the sand, would effectually stop it. The plate was put in, bolted to the one already in, and a short brace put up, temporarily, to prevent settling; but it was found that the material on which the lower end of the brace rested was insecure, and systems of A-braces, diagonals, verticals with mud-sills, etc., were tried, but nothing but failure resulted. When the heading had been extended about 12 feet and the bulkhead put in, reaching across the face of the heading from side to side and down about 6 feet, it was found that the continued settlement of the roof had reached a point when some radical change in the method must be adopted or the work would be stopped. A system of bracing was devised which supported the roof most admirably. A piece of timber 12 inches square and about 10 feet in length was so placed that one end rested against the outside of the caisson and the other extended into the face of the heading. The forward end was supported from the inside of the caisson by two iron tie-rods, 1½ inch in diameter and provided with turn-buckles, thus forming a kind of boom, which, it was expected, would sustain a load of from 50 to 75 tons. An upper member was added to this, and with a strut or two formed a sort of truss closely resembling a bridge-truss. This arrangement constituted the principal aid in completing the first section. Plate XX. is a sectional side elevation through the caisson and completed tunnel; it also shows very clearly the construction of the sides of the caisson.

The finally adopted method of building the tunnel through sand will be understood from the engravings, Plate XXII., the upper view showing the first work done upon a new section, the centre view showing the excavation about one-half finished, and the lower view representing the chamber lined with iron plates, cleaned and ready for the masonry; the respective cross-sections showing the bracing are seen in the side views.

The upper middle plate of the bulkhead of the completed section having been removed, the first crown-plate of the next section was put in and bolted to the finished ring; plates were then put at each side and bolted. The crown-plates were extended as far as possible before the next rows of plates in the heading were removed. The side plates at the crown were also carried down, so that their lower edges might be below the upper edge of the highest row in the bulkhead. Since, owing to the air-pressure, the water could not rise above the edges of these plates, the main efforts were directed toward rapidly advancing the roof-plates, in order to reach the new bulkhead. When this was reached the plates that had been taken from the rear bulkhead were put up. This formed the segment of a tube the circular portion of which was iron and the base sand. Short, thick timbers resting upon plates sunk a little distance into the sand floor supported the iron sheathing. When the excavation had been carried far enough down to admit them horizontal timbers were used as foundations for the short braces; the rear ends of these timbers were sustained by pieces placed against the finished tunnel, and the middle and forward parts by uprights on plates in the sand. The plates were slowly and carefully inserted, all digging being done with extreme caution. The excavated material was removed to the caisson-chamber, and from there taken to the surface in a manner presently to be described.

To make the joints between the plates as nearly air-tight as possible was much more essential upon this side than it had been upon the other, and it was a much more difficult undertaking owing to the total want of self-sustaining power in the sand, which, being continually saturated with water, offered but slight resistance to the passage of air. Of the many substances experimented with, silt brought from the New Jersey side proved most effectual in stopping leaks. This was plastered freely over the iron-work, and by closely watching its surface the incipient leak could be easily detected.

To insert the plates in the roof and invert was comparatively easy, but at the middle section greater caution was exercised. In order that the exposed earth might be of as small area as possible, the side and bulkhead plates were cut much smaller, as shown in plates XXIII., XXIV., and XXV. Plate XXIV. shows a ring of plates as it would appear if rolled out flat, and the following plate shows how a large plate was cut up, and also gives the dimensions of each.

As has been before stated, it was necessary to increase the air-pressure as the plates in the section were carried down; the hydrostatic head and air-pressure were kept about balanced. A pressure of from 25 to 28 pounds per square inch was required during the building of the invert.

Building bulkheads in the tunnels at certain distances and inserting air-locks had proved so satisfactory in the work at the other side of the river that, at a distance of 30 feet from the caisson, a lock 15 feet long and 6 feet in diameter was built upon a bulkhead of concrete 6 feet thick. From the bulkhead to the arch of the tunnel, and surrounding the lock, was a wall of brick masonry 3 feet thick. This lock was similar in construction to those which we have described. The principal details and location of the bulkhead and lock will be readily understood from the engravings, Plate XXVI.

The experience gained in removing the excavated material from the work upon the New Jersey side caused many changes to be made in the construction and location of the air-locks in the working-chamber of the New York caisson. The general supply-lock, previously described, had a horizontal section across its lower end, and the shaft leading to it was of ample size to admit the largest pieces of iron entering the tunnel. At each end of the cross-piece was a door, and at the junction of the shaft and cross-piece was a third door; the shaft was always open to the air. The two lower doors being closed, the brick, sand, cement, plates, etc., were lowered down the shaft by means of a block and tackle attached to a frame under an open shed built over the opening. When the air-lock had been filled the middle door was closed, and after the air-pressure in the lock equalled that in the working-chamber the end doors were thrown open and the load removed. Material to be taken out was placed in the lock, and from there raised to the surface by a hoisting-engine; it was then carried in wheelbarrows on board a scow moored to the dock. Timber and long pieces which could not be conveniently taken through this lock were admitted through the timber-lock.

A sand-pump designed by Mr. C. W. Clift, master-machinist of the work, operated most successfully, proving economical both to run and keep in repair, and effective and reliable. An almost spherical casting was formed with two necks diametrically opposite each other. A pipe leading to the suction-hose screwed into the lower neck, its upper end almost entering the upper neck. Water from a

Plate XXI.

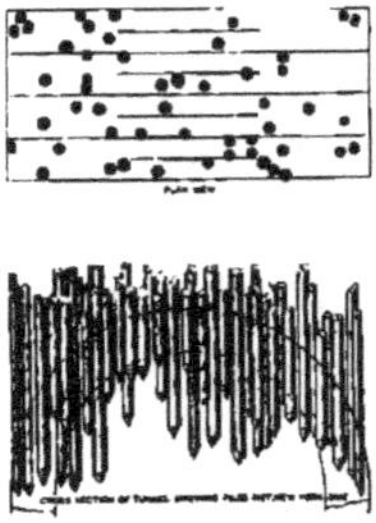

Plate XXII.

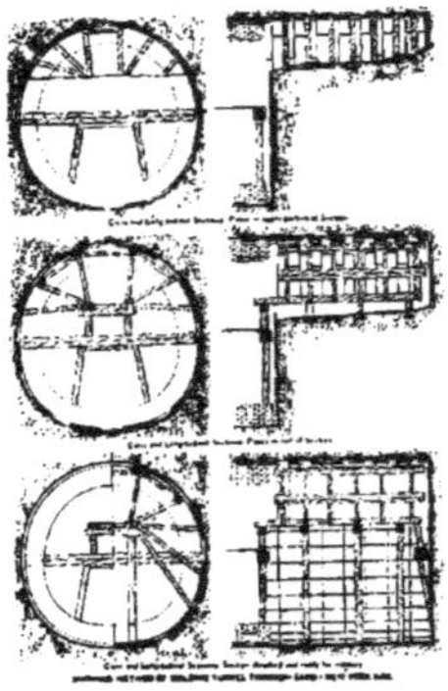

Plate XXIV.

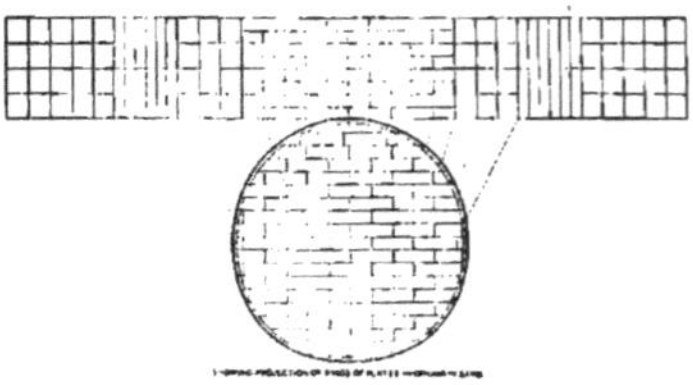

Plate XXIII.

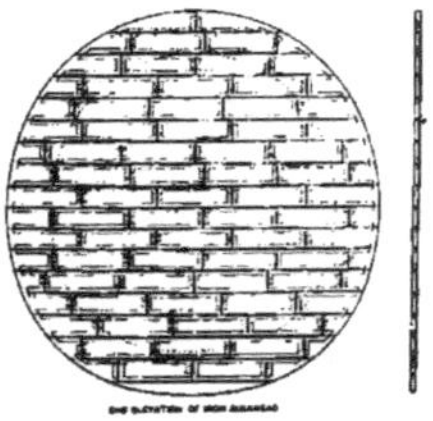

Plate XXV.

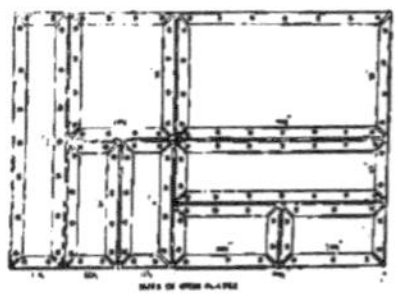

Plate XXII

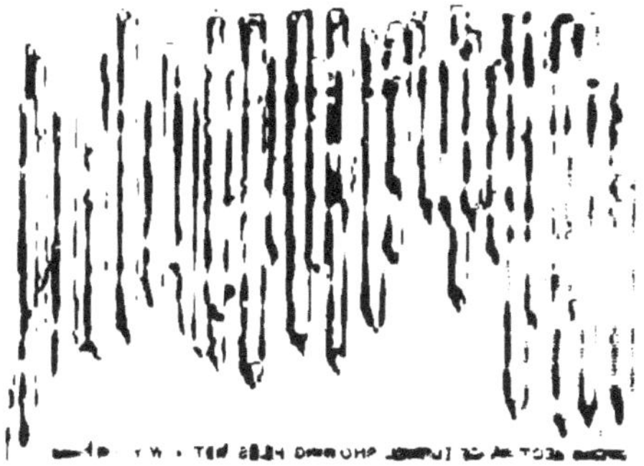

pressure-pump entered at the upper side of the globe and was conducted by a wide, curved, annular flange to the lower part of the chamber; it then turned and passed between the end of the suction-pipe and upper neck, and thence out through the delivery-pipe. The pump worked upon the principle of the induced current, and the sharp sand emerging from the suction-pipe struck against the sides of the lower end of the delivery-pipe. At that point most liable to wear was inserted a bushing of chilled cast-iron, above which was a piece of composition metal, both pieces being made thicker at the middle. These were cast to fit the pump and required no adjustment, being held in position by a screw-cap. The distance between the top of the suction-pipe and the delivery-pipe could be regulated by screwing or unscrewing the lower pipe, and any rate of discharge could be obtained with little or no trouble.

The delivery-pipe led to the surface of the ground, and the suction-pipe terminated in a flexible length which was placed in the box containing the water and sand to be removed. The pump would throw out fine sand or coarse gravel and any stone small enough to pass through the piece of chilled iron. It never clogged, and, as almost all the wear came upon the chilled-iron bushing, the cost to keep it in repair was but trifling.

The lower doors of the supply air-lock opened upon a platform built in the working-chamber and extending along the tunnels to the bulkheads; from the other side of the bulkheads the platform extended to the headings. Both tunnels were left a little more than half-full of excavated material.

On August 20, 1882, a blow-out occurred in the fifth section 65 feet from the bulkhead in the north tunnel (at that time the south tunnel had not been started). The section was 15 feet long, and had been bricked up at the invert. A plate was forced out in the iron bulkhead, allowing the air to escape. The air-pressure at the time was 26½ pounds per square inch. The plate was in the fifth row, near the end. The workmen, being warned by the peculiar hissing noise, had ample time to reach the lock, which was approached by a short flight of steps.

This extra long section—15 feet—required more than the usual time to build, so that the air, under the high pressure of 28 pounds for several days, found avenues of escape to the surface along the piles, and in time these passages became enlarged so that the plates could not be held in position, and hence control of the air was lost.

The flooded section was entered by a diver, who found the upper courses of plates in the bulkhead uninjured, but those immediately adjoining the plate at which the leak took place were much out of place. The air was then admitted, when the water was expelled from the upper part of the section, or to a level corresponding with the broken plate. This was sufficient to give a breathing-space for the men, who removed the distorted plates and inserted others. This was a labor of exceeding difficulty, prosecuted under the most trying surroundings, and which well illustrates the courage and skill of those by whom it was accomplished. It took 45 days to recover what had been lost by the blow-out.

Plate XXI. is a section and plan showing the piles met in passing under the bulkhead-wall (this drawing only represents 10 lineal feet, and as there were 25 feet in all some idea may be formed of the difficulties encountered); this is also shown in the longitudinal section, Plate XXVII. Many of these piles extended nearly through the tunnel, and they not only had to be cut off, but provision had to be made to guard against any subsequent settlement.

The masonry of the tunnel was made 2½ feet thick, and was laid in Portland cement mixed in the proportion of one cement to two sand.

During the month of May, 1883, the work was advanced at the average rate of one-third of a foot per day. This was largely due to the fact that the calking in the joints of the iron plates was forced outward by the air, and thus much time and labor were required to watch for and stop the leaks. The angle-irons were riveted flush with the four sides of the plate, and, as a consequence, the wet silt-tamping put between the flanges formed by the angles was blown outward by the air-pressure when any shifting occurred in the line of the plate; if, as sometimes happened, the plates were forced up slightly, the joint opened outward and allowed the packing to escape. This was remedied by riveting the angle-irons to the plate in such a manner as to break joint; this was accomplished by allowing two adjoining angles to project beyond the edges of the plate, while the two remaining angles were set back from the edge of the plate. By this arrangement the plate-joints and angle-joints were thrown out of line,

Plate XXVI.

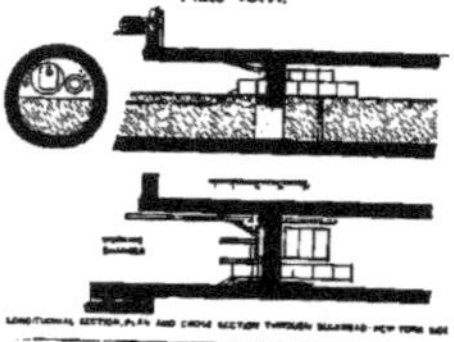

LONGITUDINAL SECTION, PLAN AND CROSS SECTION THROUGH BULKHEAD - NEW YORK SIDE

Plate XXVII.

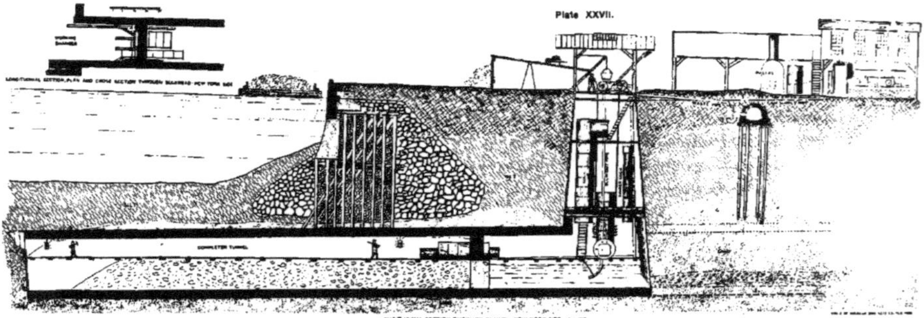

LONGITUDINAL SECTION THROUGH TUNNEL - NEW YORK SIDE.

Plate XXVI.

and any filling placed between the flanges was prevented from escaping outward by the projecting plate. After this method was adopted 1 foot of tunnel was completed each day.

About 75 feet out from the caisson had been finished when the work was stopped, November 5, 1882, for want of money, and water allowed to fill the tunnels and caisson; it remained in this condition until March 20, 1883, when steam was turned on and compressed air forced in again. In two days work was begun at the heading, and continued until July 20—four months—during which time 72 feet of masonry had been put in the north tunnel, and the south tunnel had been cut through the caisson and 23 feet of it completed. Work was again suspended because of lack of funds.

During this period better results had been accomplished than during the previous year; the men had gained experience, and all the new plans worked admirably. The average rate of progress was from 1 to 3 feet of completed tunnel per day.

When work was stopped the north tunnel had been carried so far as to be nearly out of the sand. Where the heading now rests the upper part is in silt; and as the dividing line between the silt and sand extends downward very sharply (as shown in Plate XXVII.), but a short length will have to be built when the work can be prosecuted entirely in silt. This will be a great aid to rapid work in the future, and the daily progress should equal that made at the other side of the river.

At the New York side all of the machinery was in a building located a short distance back or east of the shaft. The plant consisted of two air-compressors, two boilers, engine and dynamo, pumps, etc. The compressors forced the air into an iron reservoir, from whence it was conducted by a pipe to the tunnels; the pressure was indicated by a mercury-gauge connected with the reservoir.

The Clayton duplex and double-acting air-compressor had two steam-cylinders 22 inches in diameter, two air-cylinders 24 inches diameter, the stroke being 30 inches. It was designed to make from 70 to 80 revolutions per minute, the capacity being, by measurement, 2,500 cubic feet of free air per minute. The tops of the cross-heads are rigidly connected by a heavy rod, and the bottoms by a distance-piece which serves as a slide, working on a long, adjustable slipper-guide placed inside of the frame, and by its use the cylinders are relieved from all weight and wear excepting that caused by the piston-packing. The suction-valves are of the poppet style, and are so numerous as to give plenty of opening for admission of air until the cylinder is filled to almost the atmospheric pressure; they are placed in the air-cylinder heads, and all danger of their falling into the cylinder through the stems breaking or nuts coming off is overcome by bolts which effectually prevent the valve from falling and enable the replacing of the nut without taking off the head; or, should the stem break, the valve can be fastened against the seat until it is convenient to repair it. The discharge-valves are lifted by an adjustable tripping device which can be set to lift at any desired point in the stroke, thus affording a free escape for the air in the cylinder as soon as it has reached the working pressure. The cylinder is surrounded by a water-jacket so constructed as to compel the water to circulate from the centre along the top to and around the ends of the cylinder for one-fourth of its length, thus covering the points at which the greatest compression of air takes place, and at which the most heat is generated.

The steam and air cylinders are securely bolted to a strong bed, and are tied together by wrought-iron rods; this arrangement admits of the fly-wheel being placed in the centre of the machine, thereby giving compactness.

In the Ingersoll air compressor the steam and air cylinders were 16 inches in diameter and the stroke 24 inches; at a speed of 88 revolutions per minute the actual amount of air was 368 cubic feet per minute. The frame is rectangular in form, and is made in one casting, the vertical, parallel sides of which are stiffened by transverse ribs. The air and steam cylinders are in line, and, with two heavy fly-wheels, are supported on the frame, making it impossible for any part to get out of line. Between them is a cast-steel cross-head with swivel-block, into which piston-rods are fitted and held by a king-bolt and adjusting clamp; this construction permits the cross-head to adjust itself to any irregularity in the length of the connecting-rods. The main and cut-off valves are operated by rocker-arms placed at the rear end of the frame back of the shaft. The journal and crank-pin boxes, slides, induction and eduction valves are made of phosphor-bronze. The induction-valves are cylindrical and have their entire circumference for a wearing-surface and guide; they work in cages screwed into the cylinder-heads from the outside, and project beyond the cages sufficiently to admit springs between a rim on the valves and the face of the cages. Cylindrical balanced eduction-valves move in similar cages, and are held in place by caps bolted to the outside of the cylinder-heads, through which adjusting-screws pass. Each valve can be removed from the outside.

Cooling is effected by a small double-plunger pump, worked from the cross-head, throwing water into the cylinder. An automatic speed and pressure regulator is located on the air-cylinder and connected to the receiver by a small pipe.

All of the boilers were furnished by the Erie City Iron-Works, of Erie, Pa. There were two on the New York side, of 70 horse-power each. The dimensions were: Diameter of boiler, 60 inches; thickness of shell $\frac{11}{32}$ of an inch, of head $\frac{7}{16}$; length of flues, 14 feet; number of flues, 3 inches in diameter, 82; square feet of heating-surface, 1,050.

After mature deliberation the writer came to the conclusion that it would be better to devote a page or two to those who at various times were in charge of the tunnel rather than attempt to mention them as the description of the work progressed; it will be readily perceived that the latter plan would have involved needless repetition and annoyance.

At the beginning Col. W. H. Paine was consulting engineer. In November, 1879, while sinking the shaft, Mr. C. C. Brush, of the firm of Speilmann & Brush, accepted the position of chief engineer and continued in charge until August, 1880. Mr. E. H. Burlingame was assistant engineer and continued in charge after this period. Mr. J. F. Anderson began his connection with the work as superintendent while the bottom of the shaft was being put in, in December, 1879; he invented the pilot and remained in charge until Gen. Wm. Sooy Smith assumed control of the engineering department in

May, 1881. One year later Mr. S. H. Finch, who had been assistant engineer since February, 1880, took charge and continued until operations were suspended, Mr. C. W. Raymond being his assistant. During all the time the machinery was in charge of Mr. C. W. Clift.

The main tunnel under the river will be 5,500 feet in length, the New Jersey approach will be 4,000 feet in length and the New York about 4,500 feet, making a total of 14,000 feet, or about 800 feet over 2½ miles.

It is intended to have not less than 12 feet of silt between the top of the tunnel and water, or bed of the river, at any point.

There are about 14 cubic yards of excavation for every lineal foot of single tunnel; hence for 5,500 feet of double tunnel 154,000 cubic yards would have to be removed.

There are 3,000 bricks, 6 barrels of cement, and 1,200 pounds of iron plates per lineal foot; hence 5,500 feet of double tunnel would require 33,000,000 bricks, 66,000 barrels of cement, and 13,200,000 pounds of iron. These estimates are for the main tunnel between the shafts. The two approaches will require about double the above quantities, except in the case of the iron, the amount of which will depend largely upon the method of construction.

With the money in hand to finish the whole it is estimated that all could be completed in two and one-half years. It is to be hoped that a work of such magnitude and universally acknowledged public benefit will not long suffer for the want of the necessary means to complete it.

ROCK-DRILLS AND GENERAL MINING MACHINERY.

"CLAYTON AIR-COMPRESSORS,"

USED AND RECOMMENDED BY THE

HUDSON RIVER TUNNEL CO.

☞ Send for Catalogue.

BUILT BY THE

CLAYTON STEAM-PUMP WORKS, 45 & 47 York St., Brooklyn, N. Y.

Used in many of the principal GOLD, SILVER, COPPER, COAL, AND IRON MINES in North and South America.

SIMPLE! DURABLE!! ECONOMICAL!!!

INDEX TO ADVERTISEMENTS.

www.ingramcontent.com/pod-product-compliance
Lightning Source LLC
Chambersburg PA
CBHW032359280326
41935CB00008B/626
9783337302320